Frontiers in Clinical Drug Research-Anti Infectives

(Volume 7)

Edited by

Atta-ur-Rahman, *FRS*
Honorary Life Fellow,
Kings College,
University of Cambridge,
Cambridge,
UK

Frontiers in Clinical Drug Research-Anti Infectives

Volume # 7

Editor: Prof. Atta-ur-Rahman, *FRS*

ISSN (Online): 2352-3212

ISSN (Print): 2452-3208

ISBN (Online): 978-981-4998-09-3

ISBN (Print): 978-981-4998-10-9

ISBN (Paperback): 978-981-4998-11-6

need for a court order if at any point you breach any terms of this License Agreement. In no event will any delay or failure by Bentham Science Publishers in enforcing your compliance with this License Agreement constitute a waiver of any of its rights.

3. You acknowledge that you have read this License Agreement, and agree to be bound by its terms and conditions. To the extent that any other terms and conditions presented on any website of Bentham Science Publishers conflict with, or are inconsistent with, the terms and conditions set out in this License Agreement, you acknowledge that the terms and conditions set out in this License Agreement shall prevail.

Bentham Science Publishers Ltd.
Executive Suite Y - 2
PO Box 7917, Saif Zone
Sharjah, U.A.E.
Email: subscriptions@benthamscience.net

BENTHAM SCIENCE

CONTENTS

PREFACE

The 7th volume of Frontiers in **Clinical Drug Research – Anti Infectives** comprises five chapters that cover a variety of topics including antivirals, treatments against some gram negative and gram negative bacteria, and an overview on a few antiprotozoal drugs that target specific pathogens.

In chapter 1, Evran *et al.*, focus on aptamers with antiviral activity, as well as the use of aptamers in viral detection platforms. They also give an overview of aptamers developed against viruses, and discuss the major hurdles in aptamer use, as well as the strategies to improve the drug potential of aptamers.

In chapter 2, Leowattana *et al.* discuss host-directed, antibiotic-adjuvant combinations and antibiotic-antibiotic combination for treating Multidrug-Resistant (MDR) gram- negative pathogens (*Acinetobacter, Enterobacteriaceae, Pseudomonas,*etc.).

In chapter 3, Barbosa and Teixeira explore the current therapeutic approaches and advances in the search for alternative solutions to inhibit the opportunistic pathogen *C. difficile.*

Rivera-Fernández *et al.* in chapter 4 of the book, review the *in vitro* and *in vivo* activities of extracts, fractions, and isolated compounds obtained from different plants against *Toxoplasma gondii,* the pathogen that causes toxoplasmosis. This chapter presents information on potential leads for novel therapeutic agents for this disease.

In the last chapter of the book by Percário *et al.,* the author describe the main Amazonian species used to treat malaria and leishmaniasis in Brazilian folk medicine, relating ethnobotanical results to chemical studies, evaluation of activities, and toxicity. Several promising compounds of plants used in traditional Amazonian medicine are described

I would like to thank all the authors for their excellent contributions that will be of great interest. Also, I would like to thank the editorial staff of Bentham Science Publishers, particularly Mr. Mahmood Alam (Editorial Director) of Bentham Science Publishers, Mr. Obaid Sadiq (In-charge Books Department) and Miss Asma Ahmed (Senior Manager Publications) for their support.

Atta-ur-Rahman, *FRS*
Kings College
University of Cambridge
Cambridge
UK

List of Contributors

Antônio R. Q. Gomes
Post-Graduate Program in Pharmaceutical Innovation, Federal University of Pará, Belém, PA, Brazil

Canan Özyurt
Department of Chemistry and Chemical Processing Technologies, Lapseki Vocational School, Canakkale Onsekiz Mart University, Canakkale, Lapseki, Turkey

Elba Carrasco-Ramírez
Departamento de Microbiología y Parasitología, Facultad de Medicina, Universidad Nacional Autónoma de México (UNAM), Coyoacán, Mexico

Elisa Vega Ávila
Departamento de Ciencias de la Salud, Universidad Autónoma Metropolitana, Unidad Iztapalapa, Mexico

Ezgi Man
Department of Biochemistry, Ege University, Faculty of Science, Izmir, Turkey

Heliton P. C. Brígido
Post-Graduate Program in Pharmaceutical Innovation, Federal University of Pará, Belém, PA, Brazil

Jhony Anacleto-Santos
Doctorado en Ciencias Biológicas y de la Salud, Universidad Autónoma Metropolitana, Mexico
Departamento de Ciencias Naturales, Universidad Autónoma Metropolitana, Unidad Cuajimalpa, Mexico

Joana Barbosa
Universidade Católica Portuguesa, CBQF - Centro de Biotecnologia e Química Fina—Laboratório Associado, Escola Superior de Biotecnologia, Porto, Portugal

Juliana Correa-Barbosa3
Post-Graduate Program in Pharmaceutical Sciences, Federal University of Pará, Belém, PA, Brazil

Kelly C. O. Albuquerque
Post-Graduate Program in Biodiversity and Biotechnology of the BIONORTE Network, Federal University of Pará, Belém, PA, Brazil

Merve Gültan
Department of Biochemistry, Ege University, Faculty of Science, Izmir, Turkey

Maria Fâni Dolabela
Post-Graduate Program in Pharmaceutical Sciences, Federal University of Pará, Belém, PA, Brazil
Post-Graduate Program in Biodiversity and Biotechnology of the BIONORTE Network, Federal University of Pará, Belém, PA, Brazil

Norma Rivera-Fernández
Departamento de Microbiología y Parasitología, Facultad de Medicina, Universidad Nacional Autónoma de México (UNAM), Coyoacán, Mexico

Özge Uğurlu
Department of Biochemistry, Ege University, Faculty of Science, Izmir, Turkey
Department of Medical Services and Techniques, Hatay Vocational School of Health Services, Hatay Mustafa Kemal University, Tayfur Sökmen Campus, Alahan-Antakya/ Hatay, Turkey

Pathomthep Leowattana
Tivanon Medical Clinics, Tivanon Road, Muang, Nonthaburi, Thailand

Paula Teixeira
Universidade Católica Portuguesa, CBQF - Centro de Biotecnologia e Química Fina—Laboratório Associado, Escola Superior de Biotecnologia, Porto, Portugal

Perla Yolanda López Camacho Departamento de Ciencias Naturales, Universidad Autónoma Metropolitana, Unidad Cuajimalpa, Mexico

Ricardo Mondragón-Flores Departamento de Bioquímica, Centro de Investigación y Estudios Avanzados del Instituto Politécnico Nacional (CINVESTAV-IPN), Zacatenco, Mexico

Sandro Percário Post-Graduate Program in Biodiversity and Biotechnology of the BIONORTE Network, Federal University of Pará, Belém, PA, Brazil

Serap Evran Department of Biochemistry, Ege University, Faculty of Science, Izmir, Turkey

Tawithep Leowattana Department of Medicine, Faculty of Medicine, Srinakharinwirot University, Sukhumvit 23, Wattana District, Bangkok, Thailand

Wattana Leowattana Department of Clinical Tropical Medicine, Faculty of Tropical Medicine, Mahidol University, Rajavithi road, Rachatawee, Bangkok, Thailand

CHAPTER 1

Nucleic Acid and Peptide Aptamers as Potential Antiviral Drugs

Serap Evran[1,*], **Özge Uğurlu**[1,2], **Ezgi Man**[1], **Merve Gültan**[1] and **Canan Özyurt**[3]

[1] *Ege University, Faculty of Science, Department of Biochemistry, 35100, Izmir, Turkey*

[2] *Department of Medical Services and Techniques, Hatay Vocational School of Health Services, Hatay Mustafa Kemal University, Tayfur Sökmen Campus 31060, Alahan-Antakya/ Hatay, Turkey*

[3] *Department of Chemistry and Chemical Processing Technologies, Lapseki Vocational School, Canakkale Onsekiz Mart University, Canakkale, Lapseki, Turkey*

Abstract: Aptamers with target-specific binding properties have emerged as an alternative to antibodies. Nucleic acid aptamers are short single-stranded oligonucleotides that can fold into unique three-dimensional structures. Nucleic acid aptamers are selected from random libraries *in vitro* by using the SELEX (Systematic Evolution of Ligands by Exponential Enrichment) technology. Likewise, peptide aptamers are short peptides that can be selected *in vitro* by using different strategies including phage display, ribosome display, or mRNA display. Aptamers are superior to antibodies with regard to ease of production, high stability, small size, and low cost. Therefore, aptamers find broad use in different biotechnological and therapeutic applications. Among them, aptamer use in virus detection and antiviral therapy is one of the attractive applications. The present Covid-19 pandemic and life-threatening viral infections reveal the need for rapid therapeutic solutions that can efficiently target viral mechanisms. In this respect, the chapter is mainly focused on aptamers with antiviral activity, as well as the use of aptamers in viral detection platforms. First, we summarize aptamer selection technologies that can be performed *in vitro*. Among them, we briefly explain ribosome display, mRNA display and SELEX (Systematic Evolution of Ligands by Exponential Enrichment) technologies. Then, we review aptamers targeting viral proteins and viral invasion mechanisms. In addition, we give an overview of aptamers developed against viruses. We also discuss the major hurdles in aptamer use, as well as the strategies to improve the drug potential of aptamers.

Keywords: Antiviral aptamer, Aptasensor, Diagnostic aptamers, DNA aptamer, MRNA display, Peptide aptamer, Ribosome display, RNA aptamer, SELEX, Therapeutic aptamers.

---

[*] **Corresponding author Serap Evran:** Ege University, Faculty of Science, Department of Biochemistry, 35100, Izmir, Turkey; Tel: +90 232 3112304; Fax: +90 232 3115485; E-mail: serap.evran@ege.edu.tr

Atta-ur-Rahman, *FRS* **(Ed.)**

1. INTRODUCTION

Nucleic acid aptamers and peptide aptamers with antibody-like binding properties are promising therapeutic agents. Several aptamers are currently evaluated under clinical phase studies, but the majority of studies are focused on metabolic diseases. In this chapter, we aim to highlight the potential use of aptamers as novel antiviral agents, and their importance in diagnosis and monitoring viral infections.

The first section of the chapter gives an overview of the *in vitro* selection methods used to identify peptide and nucleic acid aptamers. Here, only some methods developed for selection from combinatorial libraries are given. This section is divided into three sub-sections as 1.1, 1.2 and 1.3 to introduce the methods of mRNA display, ribosome display and SELEX. The SELEX section is further divided into 1.4.1 and 1.4.2 to summarize two of the SELEX methods, namely cell-SELEX and bead-based SELEX. The final sub-sections 1.5 and 1.6 explore the modification strategies to improve the stability properties of aptamers for therapeutic use.

The second section of the chapter is devoted to aptamers used for the detection of viruses. The sub-sections 2.1-2.9 summarize the studies for diagnosis of Hepatitis B Virus (HBV), Hepatitis C Virus (HCV), Human Immunodeficiency Virus (HIV), Influenza, Arboviruses, SARS virus, Ebola Virus, SARS-CoV-2, and Human Papilloma Virus (HPV).

The third section of the chapter is devoted to the aptamers targeting viral proteins. Aptamers developed against several proteins of HIV, HCV, HBV, SARS, and Influenza are summarized under the sections 3.1-3.5.

1.1. The *in vitro* Selection Methods for Peptide and Nucleic Acid Aptamers

Directed evolution is a powerful tool to develop proteins with superior function and binding properties [1]. One critical step of directed evolution is the screening or selection of improved variants from a large library [2]. Advances in molecular techniques have allowed the design of highly diverse libraries of nucleic acids, peptides and antibodies [3]. To meet the demand for working with large libraries, protein display technologies have been developed for the selection, isolation and identification of proteins with the desired properties [4,5]. Display techniques are basically divided into two groups: (i) cell-free and (ii) cell-based. Cell-based approaches such as bacterial display, yeast display, mammalian cell display and phage display have some limitations regarding the efficiency of recovery and library diversity [6]. Although phage display has been widely used to select proteins and peptides with improved binding properties [7], cell-free display methods have emerged as an alternative to overcome the limitations associated with living cells [8].

Ribosome display and mRNA display are cell-free methods, which rely on *in vitro* transcription and translation of the newly formed peptide along with its encoding mRNA. Thereby, ribosome display and mRNA display can establish a direct link between phenotype and genotype. *In vitro* display and selection approach consists of 3 steps: (i) designing the initial library, (ii) performing repetitive rounds to obtain the desired characteristics, (iii) screening and characterizing the selected variants [9]. Engineered antibodies, proteins, as well as peptides for various applications in diagnostics and therapeutics have been identified by *in vitro* selection methods [10]. Ribosome display and mRNA display enable identification of high-affinity proteins or peptide aptamers [11,12], whereas SELEX (Systematic Evolution of Ligands by Exponential enrichment) method allows *in vitro* nucleic acid aptamer selection. As shown in Table **1**, those *in vitro* methods allow working with libraries of large size. Cell-based approaches are limited by cell growth and replication [13]. In contrast, *in vitro* selection methods allow precise control of many parameters like pH, temperature, buffer conditions, and ionic strength [14]. Auxiliary components including the binding target and reaction substrates can be easily added to the selection medium, thereby eliminating the toxic effect problem that may be encountered in cell-based display methods [15].

Table 1. Library size of *in vitro* selection methods.

In vitro method	Size of library	References
SELEX	10^{14}-10^{15}	[16]
mRNA display	10^{12}-10^{13}	[14]
Ribosome display	10^{13}-10^{14}	[17]

1.2. MRNA Display

mRNA display is based on the formation of a covalent link between the target peptide/protein and the mRNA encoding it. For this aim, 3' end of mRNA is modified with puromycin, an antibiotic molecule that acts like an aminoacylated tRNA. Upon translation of the modified mRNA, puromycin enters the A region of the ribosome and forms a peptide bond with the C-terminal of the polypeptide. In this way, the puromycin-modified mRNA is covalently linked to the peptide [18]. This generates a stable link between genotype and phenotype. In addition, this stable covalent bond allows selection under stringent conditions. In mRNA display method, all steps including protein expression are performed *in vitro* (Fig. 1) [19]. Cell-free translation is generally performed using eukaryotic cell lysates obtained from wheat germ or rabbit, which offer a wider post-translational modification repertoire compared to bacterial *in vivo* expression systems. For a tighter control of protein expression conditions, the PURE (protein synthesis using recombinant elements) translation system can also be used [20]. This system consists of purified components such as tRNA, ribosome, amino acids, release factors and aminoacyl tRNA synthetases required for *E. coli* translation [21].

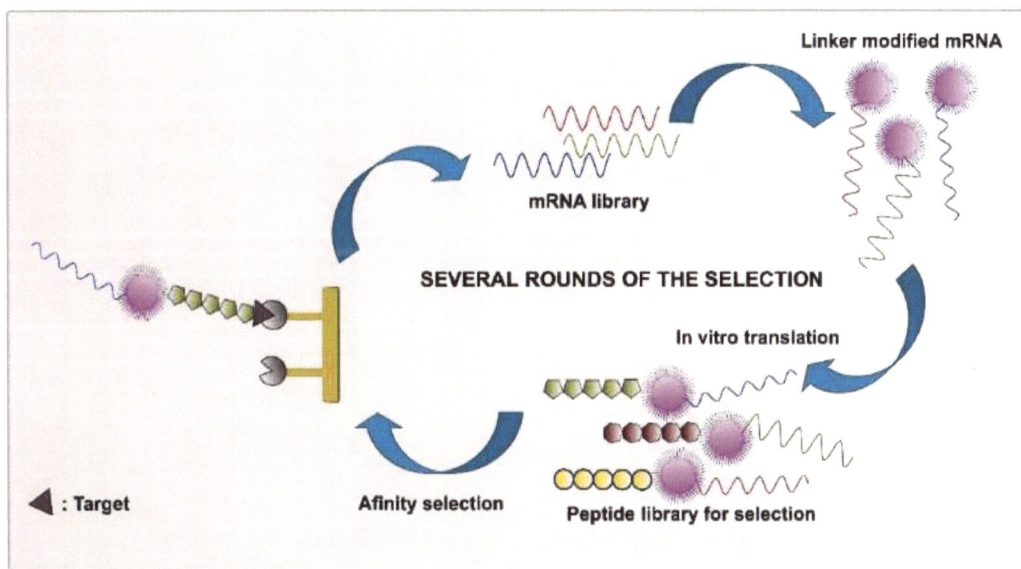

Fig. (1). Schematic representation of mRNA display.

mRNA display has several advantages over many *in vivo* and *in vitro* methods [22]. The covalent link between genotype and phenotype has a twofold benefit. Recovery of variants becomes easier, and harsh selection conditions that are not possible for many *in vivo* methods can be applied. Harsh conditions enable the development of proteins resistant to extreme pH, high temperature and high salt concentrations, which is particularly useful for enzyme engineering [23]. mRNA display has proven to be a powerful tool for peptide design, *in vitro* protein selection, studying molecular interactions, drug design, and protein engineering [24 - 27]. The developed ligands attract attention due to their superior properties and very high (nanomolar to picomolar) affinities [28].

1.3. Ribosome Display

Ribosome display has been developed as an alternative method to screen target-binding proteins/peptides against a variety of target molecules [29 - 30]. Ribosome display is an approach in which genotype and phenotype are combined through the formation of a stable triple mRNA ribosome-protein complex [31]. It enables the enrichment of high-affinity variants with the desired property from a large library through repetitive cycles of affinity-based selection and PCR amplification [32]. In this approach, transcription, translation and selection are carried out cell-free *in vitro* [18].

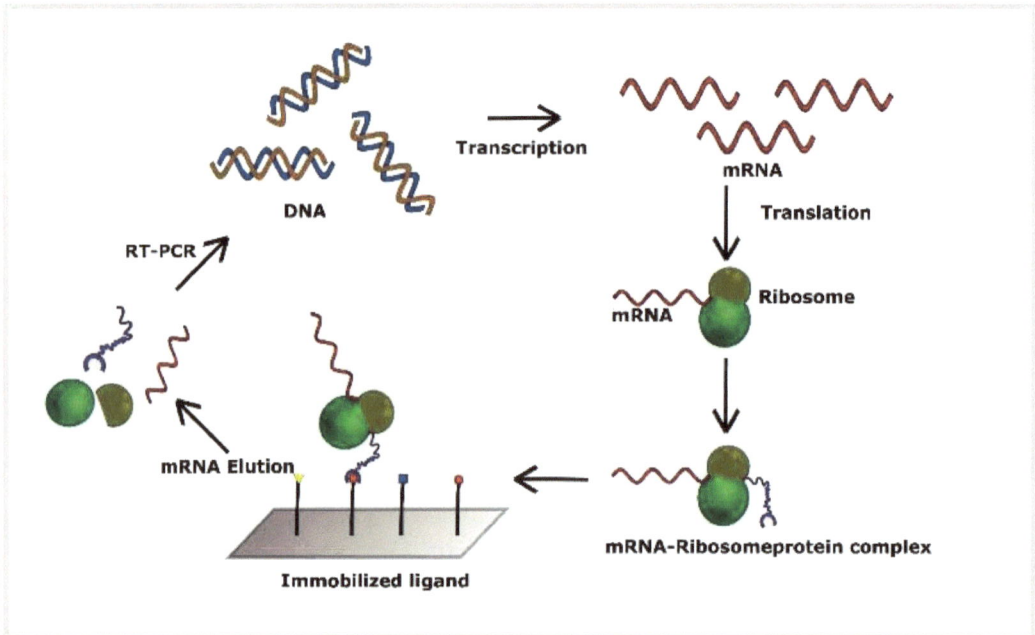

Fig. (2). Schematic representation of ribosome display.

Ribosome display basically consists of those steps; (i) preparation of DNA library, (ii) *in vitro* transcription and translation, (iii) selection and recovery of mRNA, (iv) reverse transcription and PCR step for the next round of selection. Formation of the non-covalent polypeptide-ribosome-mRNA complex is the most critical issue in this approach. The absence of a stop codon in mRNA ensures that the newly synthesized peptide and the mRNA encoding it remain attached to the ribosome Fig. (**2**) [32]. On the other hand, ribosome display suffers from nuclease-hypersensitive mRNA and low stability of the triple complex. Here, the PURE (protein synthesis using recombinant elements) system offers a solution so that the rate of mRNA recovery increases and more stable ternary complexes are produced [33].

Ribosome display is a powerful approach for high-efficiency selection of peptides and proteins. Compared to phage display, ribosome display offers a faster scanning and more efficient selection [34]. The cell-free ribosome display has been used successfully for *in vitro* selection of specific binders for more than 20 years [35]. Ribosome display is performed not only to increase the binding affinity but also the stability of proteins. For this aim, redox potential is changed by addition of dithiothreitol (DTT) during the ribosome complex generation step. Thus, correct folding of the antibodies is ensured [36]. Single-chain antibody fragments against viruses, bacteria, drugs, hormones, proteins and pesticides have been developed by using the ribosome display method [37]. This approach can also be combined with high-throughput microarray methods to add a new dimension to protein-protein interaction analysis [38]. Simultaneous scanning can be performed for various massive peptides [39]. Ribosome display has proven a useful tool in different research areas, such as diagnosis and treatment of cancer infectious diseases, autoimmune and metabolic diseases, allergic disorders, and drug design [40 - 46].

1.4. SELEX (Systematic Evolution of Ligands by Exponential Enrichment) for Nucleic Acid Aptamers

Nucleic acid aptamers are single-stranded DNA or RNA sequences that can bind to their targets due to their folded structure, which is specified by their sequences. Aptamers bind to the target molecule through hydrogen bonding, stacking of aromatic rings, van der Waals interactions, salt bridges, and electrostatic interactions [47]. The term aptamer is derived from the Latin words 'aptus' - 'fit' and 'merus' - 'particle' [48]. Aptamers are selected by *in vitro* selection - Systematic Evolution of Ligands by Exponential Enrichment SELEX [49,50]. SELEX process consists of incubation, elution and amplification steps (Fig. **3**). DNA or RNA libraries containing $\sim 10^{14}$-10^{15} different oligonucleotides are first incubated with the target molecule. Then, sequences that show no or less affinity to the target are removed from the SELEX medium by using different approaches. The sequences bound to the target are amplified and used in the next round. After the SELEX is completed, cloning and characterization steps are performed [51].

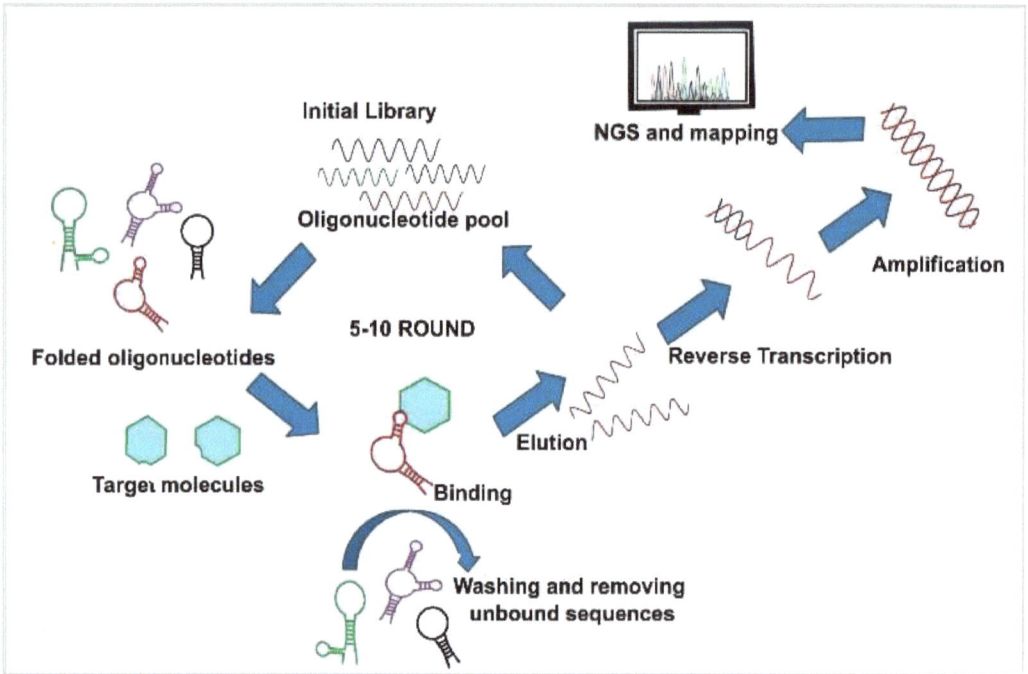

Fig. (3). Schematic representation of SELEX.

As summarized in Table **2**, aptamers have several advantages over antibodies in terms of chemical synthesis, high stability, high affinity and possibility of post-SELEX modification [52,53]. The molecular weight of the aptamers ranges from 10 to 30 kDa. With these properties, they are smaller molecules than antibodies [54]. Compared to antibodies, aptamers are highly advantageous due to low immunogenicity and cost. These properties highlight the aptamers as potential next-generation therapeutics [55,56]. Aptamers bind to a wide range of molecules, such as peptides, proteins, viruses, cells, tissues, nucleotides, antibiotics, toxins, and metal ions with high affinity and specificity [57,58]. Aptamers have been actively used in a variety of biomedical applications including molecular imaging, treatment and diagnosis of diseases, detection of biomarkers, and targeted drug delivery [59 - 67]. In addition, aptamers can be used in chromatography [68], Western-blot [69], surface plasmon resonance (SPR) [70], biosensors and micro-arrays [71]. Since the first introduction of SELEX technology, several protocols including bead-based SELEX, microarray-based SELEX, capture-SELEX, capillary electrophoresis SELEX, microfluidic SELEX, *in silico* SELEX, *in vivo* SELEX, and cell-SELEX have been proposed [16,72].

Table 2. Comparison of aptamers and antibodies

Feature	Aptamers	Antibodies
Size	6- 30 kDa	150- 180 kDa
Manufacture	*in vitro* SELEX method Chemical synthesis 2-8 weeks	*in vivo* Biological system 6 months
Stability Temperature	Long shelf-life, very stable Highly resistant (up to 95° C)	limited shelf-life Highly susceptible
Modification	Convenient	Restricted
Batch to batch variation	Low	Significant
Potential Targets	Wide range: small molecules to large molecules, ions, toxins, viral particles	immunogenic molecules

1.4.1. Cell-SELEX

Unlike conventional methods based on known targets, cell-SELEX does not require prior knowledge about the target protein. In the classical approach, target protein is obtained from living cells by overexpression and purification. However, these processes are costly and time-consuming. In addition, purified proteins may be less stable than their natural forms. Possible conformational differences in the target protein can adversely affect aptamer selection [73]. In cell-SELEX, target protein in the living cell preserves its natural conformation. Thus, the selected aptamers can recognize their target in its natural conformation. Cell-SELEX can be applied against virtually any cell line, parasite, virus and bacteria [74]. Since target molecules are in their active and natural form in the cell, the selected aptamers can be used directly for cellular imaging, diagnosis and therapy [75]. Cell-SELEX enables development of high affinity aptamers against specific molecules in the target cell. These aptamers are capable of separating target cells from other cells, which could be useful for cell profiling, cell capture, cell and tissue imaging, cell detection, drug delivery, diagnosis and treatment [76,77].

Although cell-SELEX consists of similar incubation, elution and amplification steps, it requires more cycles and longer time compared to the classical approach [78]. Up to 35 rounds may be required to obtain high affinity sequences. In addition, any damage to target cell can affect selection results [79]. Cell number, initial library design, separation method, the number of selection rounds and temperature are the critical factors. The interference of dead cells is one important problem in cell-SELEX. Non-specific binding of aptamer library to dead cells adversely affects aptamer selection. In order to avoid this problem and remove dead cells from the environment, fluorescence activated cell sorter (FACS) and microbead-based methods have been developed [80,81]. Another critical factor is temperature. Cell-SELEX is usually carried out at 4°C, but binding characteristics of aptamers may differ at the physiological temperature [82].

1.4.2. Bead-Based SELEX

One of the commonly used strategies in SELEX is immobilizing the target molecule on a solid support material of sepharose, agarose or sephadex. After the protein-bound beads are incubated with initial library, aptamer-target protein complex can be easily separated [83,84]. Bead-based SELEX is advantageous in many ways. It can be easily adapted to different targets such as peptides, proteins and small molecules. Aptamer selection is much faster and more practical than conventional SELEX methods. Selection conditions can be changed easily [85].

1.5. Post-SELEX Modifications

Despite the great potential of aptamers as therapeutics, attack by nucleases and renal filtration can limit their use. Post-SELEX modification to increase stability and affinity of aptamers is a critical stage to avoid these possible risks. As one of the post-SELEX strategies, truncation allows removing nucleotides that are not involved in target binding, which may result in enhanced affinity [86 - 88].

Aptamers can be chemically modified to impart nuclease stability. Sugar rings, bases, and linkages are the best candidates for different modifications [89]. Modification of oligonucleotides according to their mirror image (Spiegelmers) is a straightforward approach. Usually, nucleases recognize the D-form (native) but not the L-form. L-aptamers can bind their targets with high specificity like D-aptamers, but do not activate the immune system. Epitope-selective Spiegelmers developed against troponin complex subunits [90] and L-aptamer targeting D-vasopressin [91] are among the successful examples. In addition, 2'-Amino modification [92,93] and 2'-Fluoro modification are common post-SELEX strategies [94,95]. It has been shown that non-specific interaction tendencies of aptamers for human neutrophil elastase can be reduced by 2'-F purine modification [94]. This study shows that 2'-modifications provide more than increasing thermal stability and nuclease resistance. Similarly, $2'-O-CH_3$ modification has been shown to improve several properties, such as nuclease resistance and thermal stability [96,97]. As another post-SELEX strategy, locked nucleic acid (LNA) modifications can also be introduced into aptamers. In a study targeting the trans-activating responsive (TAR) element of HIV-1 genome, 2'--methyl aptamer containing two LNA residues in the loop has been shown to inhibit TAR-dependent transcription [98]. 3' or 5'-end-capping of aptamers with biotin, streptavidin, cholesterol, PEG, proteins, amine, and phosphate are other modifications for biosensor and therapeutic applications [99,100].

Most aptamers in clinical trials contain a thiophosphate backbone, which is responsible for reduced thermal stability. Besides, aptamers with thiophosphate substitutions may show more nonspecific interactions than unmodified aptamers [101,102]. Phosphodiester bonds can be replaced with methylphosphonate or phosphorothioate linkage [103]. As shown for α thrombin-PS2 RNA aptamer with phosphorodiothionate (PS2) linkages, the binding affinity can be increased dramatically (from nanomolar to picomolar range) [104].

SOMAmers (Slow off-rate modified aptamers) have become popular due to their superior affinity and kinetic properties, as well as nuclease resistance [105]. In a very recent study, SOMAmers have been used for proteomic screening [106]. SOMAmers have also been utilized successfully for isolation of cells from the complex medium [107]. Such studies reveal the high potential of SOMAmers for extraction and detection of viral factors and viruses from real human samples.

1.6. Modification of Peptide Aptamers for Therapeutic Applications

Low stability of peptide aptamers is one of the most important problems in therapeutic applications. As a result of protease degradation, the half-life may decrease to minutes. Peptide aptamers, like nucleic acid aptamers, can acquire many different improved properties with modifications [108]. Modification of the peptide backbone by different strategies, such as cyclization, mirror image transformation, retro-inversion, and peptidomimetics usage is a useful approach to increase peptide stability [109,110]. It has been suggested that the peptide synthesized in the reverse order of the natural peptide would have a side chain structure and target binding affinity similar to the parent peptide and impart resistance to proteases. This strategy has been successfully used for Prosaptide peptides to improve bioactivity [111,112]. In addition to these modifications, conjugating peptide aptamers with functional proteins is also useful. For this aim, it is possible to design genetic fusion proteins and make different modifications according to the purpose.

2. APTAMERS IN VIRAL DIAGNOSIS

Viruses are agents that multiply in living cells of host organisms and cause infection. When viruses enter the cell, they can stay in the dormant phase for a long time or they can start the infection immediately. There are several severe diseases that are caused by viruses such as Covid-19, SARS MERS, Spanish flu, hepatitis, cervical cancer, influenza, and human immunodeficiency virus (HIV). Today, the new type of coronavirus is still a major health threat. Rapid spread of Covid-19 in a short time around the world, reveals the urgent need for fast, cheap and sensitive viral diagnostic platforms. Detection methods for diagnosis of viral infections can be generally divided as gene-based or antibody-based approaches. Gene-based methods rely on polymerase chain reaction (PCR), including real-time PCR, LAMP (loop mediated isothermal amplification-based assay), NASBA (nucleic acid sequencing-based amplification). Antibody-based approaches are used to detect viral proteins and host antibodies [113]. As summarized in (Table 3), each approach has some advantages and disadvantages.

Table 3. Advantages and disadvantages of methods for viral infection diagnosis.

Method	Target	Pros	Cons	Ref.
PCR based methods	Nucleic acid	Low detection limit Sensitive	Specialized equipments, need experts, time- consuming, contamination, false positive	[114]
Antibody based methods	IgG, IgM	Rapid, cheap, no need experts, portable	Post-infection, not specific	[115]
Viral culture	Virus	Specific	Time-consuming (6-10 days), low sensitivity, safety	[116]
Antigen Tests	Viral antigen	Portable, point-of-care, sensitive, early detection is possible	Sensitivity changes with commercial assay	[115]

In general, current diagnostic tools are highly expensive and time-consuming. For that reason, aptamers with superior properties have great potential to be used in diagnostic tools. Aptamers can chemically synthesized and easily modified with biotin, polyethylene glycol (PEG) or fluorescent dyes. Moreover, unlike antibodies, they do not show batch-to-batch variations.

2.1. Aptamers in Diagnosis of Hepatitis B Virus (HBV)

HBV, a member of *Hepadnaviridae* family is a partially double-stranded DNA virus. HBV genome is composed of ~3200 base pairs. Hepatitis B virus surface antigen (HBsAg) is the most studied and used antigen in diagnosis of HBV in blood samples [117]. HBsAg is the first virological marker to appear in blood circulation [118]. Recombinant HBsAg is used to induce neutralizing antibodies for vaccination [119]. Despite the effective vaccination, HBV is still the cause of deaths [119]. In hospitals, enzyme-linked immunosorbent assay (ELISA) kits are commonly used to detect HBsAg. Conventional methods detect HBsAg 6-8 weeks after the infection [118]. Therefore, new sensitive and fast methods are urgently needed for detection of HBV. Xi *et al.* [118] obtained three aptamers against HBsAg by using SELEX. The selected aptamers were immobilized onto the carboxylated magnetic nanoparticles to develop a chemiluminescence aptasensor platform. Detection limit of the aptasensor was found 0.1 ng/mL in serum samples, which was five times lower than the limit of enzyme-linked immunosorbent assay (ELISA). Additionally, the authors tested the serum samples from hepatitis A and hepatitis C patients, as well as a mixed sample of hepatitis A, B and C to evaluate the specificity of selected aptamer. The authors found that hepatitis A and C did not interfere with the detection. Mohsin *et al.* [120] developed a highly sensitive electrochemical aptasensor to detect HBsAg. They used the previously developed aptamer [118] and increased the detection limit to 0.0014 fg/mL by using the carbon electrode modified with gold nanoparticles [120]. Also, the aptasensor allowed detection of HBsAg in the range of 0.5-2.0 fg/mL with good recovery in real human serum samples. The selectivity of aptasensor was tested with prostate-specific antigen, vitamin C, glucose and fetal bovine serum.

The e antigen of hepatitis B (HBeAg) is released during hepatitis B virus replication. For that reason, HBeAg is a good marker of chronic infection and related with the risk of developing liver cancer and cirrhosis [121]. Although some treatment options are available, they can only suppress the replication of virus [119]. Therefore, monitoring the levels of HBeAg in serum has crucial importance to follow the success of therapy. Considering the increase in number of chronic HBV patients, there is a great demand for cheap, fast and sensitive detection methods for HBeAg. In order to detect HBeAg in serum samples, Liu *et al.* [121] developed a sandwich assay based on reporter and capture DNA aptamers. For this aim, the authors modified the previously developed aptamers [122]. The reporter aptamer was truncated to 40-mer from original 80-mer and modified with a G-quadruplex and two loops, which yielded a K_d value of 0.4 nM. Similar truncation and modification studies were performed on the capture aptamer, which displayed a K_d value of 1.2 nM. The sandwich assay achieved detection of HBeAg in the range of 0.1−60 ng/mL in real HBV serum samples.

2.2. Aptamers in Diagnosis of Hepatitis C Virus (HCV)

HCV belongs to the family of *Flaviviridae*. It is small, positive-sense single-stranded, enveloped RNA virus, which is known to cause acute liver disease. HCV contains a small genome of 9600 base pairs. With early diagnosis, the success of anti-HCV treatment can reach up to 98% [123].

Serological tests based on anti-HCV antibodies and molecular tests based on detection of HCV RNA are commonly used. However, these are expensive techniques that need special facilities and experts. In addition, these tests are not appropriate for patients with suppressed immune system and they can not identify HCV in the early phase. ELISA assays can achieve a detection limit of 10^{-14} M [124]. Aptamer-based biosensor platforms are promising for early HCV diagnosis. As summarized in Table **4**, DNA or RNA aptamers were developed against different antigens.

Table 4. Summary of DNA and RNA aptamer which can be used in diagnosis of HCV.

Aptamer	Target	Reference
RNA	HCV core antigen	[125]
DNA	HCV core antigen	[127]
DNA (Bz-Du modified)	E2	[128]
DNA	Envelope Glycoprotein E2	[129]
RNA	HCV helicase	[130]

In aptasensor studies, HCVcoreAg with a conserved protein structure was targeted because of the link between HCVcoreAg concentration and viral RNA [125 - 128]. The first aptamers to recognize HCV antigen in serum samples were developed by Lee *et al.* [125]. RNA aptamers were selected after nine SELEX rounds. The aptamers showed high affinity to HCV core antigen, but not to NS5 antigen. Due to their smaller size, aptamers can overcome the problems encountered in design of antibody-based nanowire sensors [123]. By taking this advantage, Malsagova *et al.* [123] reported a novel nanowire aptamer-sensitized biosensor to detect HCVcoreAg in serum samples. The authors used the previously developed DNA aptamer against HCVcoreAg [127]. They designed an aptasensor with a detection limit of 10^{-15} M capable of working in acidic and neutral buffer. Also, the aptasensor showed good reproducibility in real serum sample

Core protein is found in both infectious and non-infectious phases of HCV, but E2 protein is generally related with the infectious dose of HCV [128]. For that reason detection of E2 is valuable for diagnosis. Park *et al.* [128] performed SELEX by using 5-benzylaminocarbonyl-dUridine (Bz-dU) instead of thymine and obtained four aptamers with K_d values ranging between 0.8–4 nM. Then, the authors developed an Enzyme Linked Aptasorbent Assay (ELASA) to detect HCV E2 protein. The assay was based on using two aptamers for capture and detection purposes.

2.3. Aptamers in Diagnosis of Human Immunodeficiency Virus (HIV)

HIV is a member of lentivirus and a part of the retrovirus subgroup [131]. HIV affects the immune system by damaging $CD4^+$ T cells and causes acquired immune deficiency syndrome (AIDS). Antiretroviral drugs are useful to prolong the life of patients with HIV but they cannot eliminate the HIV viruses from the body. HIV enters into cells through interaction between viral surface protein gp120 and host CD4 [132]. Enzyme-linked immunosorbent assay (ELISA) and real-time PCR are used to detect HIV infection in clinical samples. Combination of antigen/antibody immunoassays usually detects the HIV-1 and HIV-2 antibodies and p24 antigen of HIV-1 [132]. The detection range of commercial ELISA assays is 0.2–10 pg/mL for HIV-gp24 glycoprotein capsid antigen [133].

The vast majority of aptamer studies were focused on the therapeutic use of aptamers. Some studies showed that the obtained aptamers can also be used as diagnostic tools. Generally, aptamers were developed to recognize the viral transactivator (Tat), Rev and gp120 proteins [61]. Tat and Rev proteins are vital for viral replication [134]. Since Tat protein is an early sign of HIV exposure, it is useful for early detection of HIV, as well as it is a target for antiretrovirals [61]. Çağlayan and Üstündağ [133] developed a surface plasmon resonance enhanced total internal reflection ellipsometry (SPReTIRE) technique to detect Tat protein. The authors used 5 different antiTAT aptamers to detect TAT in the range of 1 nM −500 nM. Tombelli *et al.* [135] developed aptamer-based surface plasmon resonance (SPR) and quartz crystal microbalance (QCM) biosensors to detect Tat protein. High selectivity and sensitivity were obtained for both aptamer-based sensor platforms. The linear range of TAT protein was 0–1.25 and 0–2.5 ppm for QCM and SPR, respectively. Yamamoto *et al.* [136] obtained aptamer-derived oligomers against Tat protein. They designed the new aptamer by splitting two oligomers of RNA^{Tat}. One of them had a hairpin structure labeled with 5'-fuorophore and 3'-quencher. The other oligomer was nonstructured. The assay was based on the fluorescence signal in the presence of Tat. Fatin *et al.* [131] developed a split RNA aptamer-based method to detect Tat by using unmodified gold nanoparticle (GNP)-based colorimetric assay, which yielded a detection limit of 10 nM.

CD4 receptor is the primary receptor for HIV [61]. Monoclonal antibodies are used for blocking HIV entry to the cells [137]. Kenneth *et al.* [138] developed a RNA aptamer against recombinant CD4, and then the fluorophore labeled aptamers were used for the flow cytometry analysis.

2.4. Aptamers in Diagnosis of Influenza

Influenza is the most common infectious disease in worldwide. World Health Organization (WHO) estimates that nearly 1 billion people get influenza annually and it causes approximately 300,000-500,000 deaths. Influenza viruses belong to *Orthomyxoviridae* family, and their genome is negative sense single-strand segmented enveloped RNA. There are four types of influenza viruses, namely A, B, C and D. Influenza A and B are the main cause of human respiratory disease. A and B types have eight RNA segments, while B and C have seven RNA segments [113]. Haemagglutinin (HA) and neuraminidase (NA) are viral proteins that show high antigenic variety. Because of this property fast diagnostic tools could be developed for detection of influenza. As shown in Table **5**, nucleic acids aptamers hold great promise as biorecognition elements.

Table 5. Nucleic acid aptamers with potential use in diagnosis of Influenza.

Target	Aptamer	K_D	Reference
Mini HA 4900 and H1N1 virus	ssDNA	19.2 nM	[147]
Recombinant HA (H5N1 and H7N7)	RNA	170 pM	[148]
Whole HA (H3N2)	RNA	188 pM	[149]
Whole HA (H1N1)	RNA	67 fM	[150]
Recombinant HA (H5N1)	ssDNA	4.65 nM	[151]
HA proteins (H1N1, H5N1, H3N2)	ssDNA	15.3 nM	[141]
Whole H5NX (H5N1, H5N8, H5N2)	ssDNA	8×10^4 to 1×10^4 EID50/mL	[152]
Whole H1N1	ssDNA	55.14 ± 22.40 nM	[153]
Whole H5N2	ssDNA	6.913×10^5 EID50/mL to to 1.27×10^6 EID50/mL	[154]

Kukushkin *et al* [139] reported a sandwich surface-enhanced raman scattering (SERS) assay against hemagglutinin by using the previously developed DNA aptamer RHA0385 capable of recognizing the H1N1, H3N2, and H5N1 strains of influenza [140]. Detection limit of aptamer-based SERS assay was 10^4 virus particles per sample. This assay was found much more sensitive and faster than conventional methods. Bai *et al.* [141] developed an electrochemical impedance spectroscopy (EIS) aptasensor platform for diagnosis of H1N1, a subgroup of influenza B. They developed aptamers against the inactivated form of H1N1 to reduce infection risk during the experiments. The EIS aptasensor demonstrated high selectivity and sensivity to H1N1 with a limit of detection value (LOD) of 0.9 pg/ mL. Moreover, the aptasensor showed selectivity towards inactivated H1N1.

In another study, DNA aptamer with a K_d value of 77.6 ± 5.6 nM was selected against the nucleoprotein of influenza [142]. In this study, a new strategy was used and aptamer pool was incubated with antigen-antibody complex to eliminate the sequences binding to the antigen- antibody binding site.

Avian influenza (AI) is a major threat for poultry industry. There are also some reported infections in humans. Avian influenza subgroups such as H5 and H7 are of concern because of their potential mutation and transmission to humans [143]. Currently, culturing, detection of viral RNA and antigen tests are used for the detection. However, these methods are highly time-consuming. The two aptamers with potential use in point-of-care detection systems were found highly specific to H5N. Kwon *et al.* [144] reported a field-effect transistor (FET) label-free assay by using the aptamer against hemagglutinin (HA) protein. Detection limit of the aptamer-based assay was found 5.9 pM.

Chen *et al.* [145] developed a surface-enhanced raman scattering (SERS) aptasensor to detect influenza A/H1N1 virus. Detection limit of the assay was calculated as 97 PFU mL^{-1} and detection time was approximately 20 min. The test was not affected by bovine serum albumin (BSA), mucin and serum.

Kang *et al.* [146] brought a different perceptive to detect influenza A or B virus. In order to amplify the signal, the authors built a sandwich DNA aptamer by using streptavidin-fused replication protein A conjugated with biotin-horseradish peroxidase. The assay was successfully used samples from patients.

2.5. Aptamers in Diagnosis of Arboviruses

Arboviruses or Arthropod-borne viruses cause disease in humans. *Bunyaviridae, Flaviviridae*, and *Togaviridae* families of arboviruses share similar transmission pathway [155]. Arboviruses circulate in wild animals, then transmit to the arthropods, such as tick and mosquitoes. Humans become infected by bite of virus-carrying insects. Arboviruses include many viruses such as Dengue (DENV), chikungunya (CHIKV), West Nile virus (WNV), Yellow Fever, Rift Valley fever, Thick Borne encephalitis virus (TBEV), Zika virus and Crimean-Congo hemorrhagic fever (CCHF).

Dengue virus, zika and yellow fever belong to *Flaviviridae* family [156]. DENV infects 390 million people annually [157] *Aedes aegypti* is the principal mosquito vector of DENV [158]. Basso *et al.* [157] developed aptamer-conjugated hybrid nanoparticles to detect four types of Dengue virus. The conjugated aptamer did not bind to the yellow fever and zika virus.

Lee and Zheng [159] developed a sandwich ELISA method to detect NS1 protein of zika virus. Two aptamers (aptamer 2 and aptamer 10) with the highest affinity to NS1 were selected after 7 rounds. The authors tried aptamer 2/aptamer 10 and aptamer 2/antibody combinations to detect NS1. In both combinations, Aptamer 2 (truncated 41-mer, K_D= 45 pM) was used to capture the target protein and antibody was used to detect NS1. In the second assay, aptamer 10 was biotinylated at its 5′ end and combination of aptamer 2/aptamer 10 achieved detection of NS1 in the range of 100 ng/mL^{-1} μg/mL. However, aptamer/ antibody combination was found more successful with a detection range of 0.1−1 ng/mL in buffer, 1−10 ng/mL in 10% human serum and >10 ng/mL in 100% human serum. In another study, Alves *et al.* [160] reported DNA aptamers against ZIKV NS5 protein. Bruno *et al.* [171] developed DNA aptamers against different virus proteins and viral particles (Table **6**).

Table 6. Viral particles which were targeted by Bruno *et al.* [155].

Virus	Target
CCHF	Altamura Gn 611 11E7a, 11E7b, 11E7c
Chikungunya	E1a Peptide
Dengue Type 1,2,3,4	Recombinant E antigen
TBEV	CE/gE
WNV	E protein

Rift Valley fever virus is a mosquito borne virus and has negative single-strand enveloped RNA genome. It causes hepatitis, hemorrhagic fever and in humans [161]. In the study of Ellenbecker *et al.* [161] RNA aptamers against the nucleocapsid protein of Rift Valley fever virus (RVFV) were developed. Kondratov *et al.* [162] obtained DNA aptamers after 15 SELEX rounds for the surface protein of TBEV virus. Saraf *et al.* [163] reported a microfluidic detection system based on aptamers. Two aptamers for CHIKV E1 and ZIKV envelope proteins were used for capture and detection. The assay successfully detected the CHIKC and ZIKV among other arboviruses in PBS and 10% diluted defibrinated calf blood, with detection limits as low as 1 pM and 10 pM, respectively.

2.6. Aptamers in Diagnosis of SARS Virus

Severe acute respiratory syndrome (SARS) was responsible of nearly 800 deaths during SARS outbreak between 2002 and 2003 [164]. SARS coronavirus is positive single-stranded RNA virus. Nucleocapsid (N) protein of SARS is a diagnostic marker. In addition to molecular methods, ELISA can be used to detect antibodies against N protein, but this method is not suitable for early detection because antibody production takes 2-3 weeks after infection [165]. To overcome this problem, Ahn *et al.* [165] developed high-affinity RNA aptamers against N protein. In addition, the authors verified the interaction between RNA aptamer and N protein by using electrophoretic mobility shift assay (EMSA). Furthermore, they used this aptamer to capture N protein in a nanoarray chip and achieved a detection limit of as low as 2 pm/mL (42 fM). In another study, DNA aptamer with a dissociation constant of 4.93± 0.30 nM was developed to recognize N protein [166].

2.7. Aptamers in Diagnosis of Ebola Virus

Ebola virus is a highly pathogenic virus that belongs to *Filoviridae* family [167]. Early detection of Ebola is essential to control the infection. Until now, there are six types of known ebolaviruses, with four of them causing severe or deadly hemorrhagic fever in humans [132]. Zhan *et al.* [167] developed highly specific DNA aptamers for GP and NP proteins with dissociation constants of 4.1 ± 0.9 and 8.1 ± 2.4 nM, respectively. The selected aptamer for GP achieved a detection limit of 4.2 ng/mL.

2.8. Aptamer in Diagnosis of SARS-COV-2

Coronaviruses are single-stranded RNA viruses from *Coronaviridae* family. In December 2019, the new type of coronavirus caused a global pandemic. The pandemic reveals the importance of fast and sensitive methods for detection of SARS-CoV-2 (Table 7). Currently, the most common method is reverse transcription polymerase chain reaction (RT-PCR). However, nasal swab samples may not contain sufficient genetic material for the analysis. In addition, RT-PCR is time-consuming and can sometimes give false-negative results. For that reason, computed tomography (CT) scans are combined with RT-PCR for diagnosis of SARS-CoV-2. In addition to CT and RT-PCR, antibody tests are also used to detect SARS-CoV-2, but they are not effective for early detection of infection because production of IgM and IgG takes time [168].

Table 7. Detection methods for SARS-CoV-2.

Method	Target	Advantages	Disadvantages	Ref.
Real time RT-PCR	Viral RNA	Early detection, Thousands of samples in one day, Sensitivity 95%	Cross reactivity with other species	[169]
RT-LAMP Colorimetric LAMP	ORF1ab gene and S gene, Conserved region for nucleocapsid protein, Viral RNA among RdRp gene SARS-CoV-2 RNA	Fast Rapid diagnosis Medium cost	Requires validation Accuracy need to be determined	[170 - 174]
CRISPR/Cas13a	Viral RNA	Rapid, low cost	Validation	[175]
Enzyme immunoassay (EIA)	SARS-CoV-2 specific antibodies, antigens	Information about infection status and SARS-CoV-2 exposure	Time-consuming, antigen detection is less accurate than RT-PCR	[176]
Rapid tests for SARS-CoV-2	Antibodies/antigens	Rapid (15 min)	Low accuracy	[177]
Serum virus neutralization assay (svn)	Antibodies	High accuracy	Long analysis time, high cost	[170]

In 2003, during the SARS outbreak, nucleocapsid (N) protein was shown to be significantly important as a diagnostic biomarker [178]. In the early days of infection, detection of N protein is more sensitive (90% positive) compared to detection of antibodies (21.4%) and viral genome (42.9%) [179]. Another important advantage of N protein is it can be detected in urine, fecal and nasopharyngeal aspirate [180]. Furthermore, N protein is one of the most abundant proteins in SARS-CoV-2. Other valuable feature of N protein is that it has low similarity between other human coronaviruses except SARS-CoV, eliminating false-positive results. N proteins from SARS-CoV and SARS-CoV-2 share 91% sequence homology [181]. Cho *et al.* [166] developed DNA aptamer against N protein of SARS coronavirus. Then, the aptamer was modified to generate three different variants, which showed similar affinity to N protein of SARS-CoV-2 and a detection limit of 10 ng/mL was achieved [181].

Zhang *et al.* [168] developed DNA aptamers against N protein of SARS-CoV-2 after 5 selection rounds. During SELEX, they added serum to the selection buffer to improve specificity of aptamers. The initial sequence length of aptamers was 76mer. Aptamer Np-A48 was truncated and showed the best binding affinity with a K_d value of 0.49 nM. All four aptamers recognized Np with high affinity, and their affinity was below 5 nM. The aptamers were used in gold nanoparticle immunochromatographic strip (GIS) and ELISA to detect N protein in serum and urine. The authors showed that aptamer- antibody combination gave the best result.

Song *et al.* [182] developed DNA aptamers for receptor binding domain (RBD) of SARS-CoV-2 by employing ACE2 receptor during SELEX and a machine learning program. The aptamers were selected after 12 SELEX rounds. K_d values of the selected aptamers CoV2-RBD-1 (51-mer) and CoV2-RBD-4 (67-mer) were 5.8 nM and 19.9 nM, respectively. Furthermore, stimulations and experiments showed that aptamers were competed with ACE2 to bind RBD.

2.9. Aptamers in Diagnosis of Human Papilloma Virus (HPV)

Human papilloma virus (HPVs) is a non-enveloped, double-stranded small DNA virus, which belongs to the *Papillomaviridae* family. HPVs, which are transmitted most commonly by sexual contact infect squamous epithelia and cause warts or precancerous lesions. Most HPV infections are resolved by the immune system in 1-2 years after infection, but persistent infection emerges in nearly 2% of the infected people. HPV-16 is the most common class of HPV, which is related with cervical cancer. In addition, HPV is found approximately 80% of oropharyngeal cancers [183]. E6 and E7 oncoproteins are good targets for detection of HPV. E6 protein of HPV-16 binds to p53 [183] and E7 protein, an acidic phosphoprotein, binds to pRb and other related proteins [184]. Toscano-Garibay *et al.* [184] developed an RNA aptamer named G5a3N.4, which showed a K_d value of 1.9 µM for the E7 protein. In another study, reduced graphene oxide field effect transistors (rGO-FETs) were developed to detect the E7 protein and this method allowed detection of E7 up to 100 pg/mL (1.75 nM) [185]. In addition, a peptide aptamer was developed against E6 and used to understand apoptotic role of E6 in HPV positive cells [186].

2.10. Aptamer in Diagnosis of Herpes Simplex Virus (HSV)

Herpes simplex virus is a member of *Herpesviridae* family and identified as two types, namely HSV-1 and HSV-2. HSV-1 generally causes cold sores by oral-oral contact but can sometimes produce genital herpes. HSV-2 is the main reason of genital herpes with transmission by sexual contact. These infections are lifelong and primarily infect epithelial tissues, then invade the nervous system. HSV-1 and HSV-2 contain high homology (range of 50- 70%) [187].

In the first step of infection, glycoprotein C and B interact with the proteins on the host cell. Then, gD, gB, gH and gL play important roles for entry into the host cell. In particular, gD recognizes two human cell receptors, which are herpesvirus entry mediator (HVEM) and nectin-1. gD proteins from HSV-1 and HSV-2 show high homology.

Gobinath *et al.* [187] developed RNA aptamers against the gD protein of HSV-1. They selected aptamer 1 and 5, which showed high affinity to gD with K_d values of 109 and 39 nM, respectively. The aptamers were able to differentiate between the gD proteins of HSV-1 and HSV-2. In the study, 2'-fluoro modification was performed to make aptamers more resistant to nuclease attacks.

Human herpex simplex virus 5 (HHV-5), also known as cytomegalovirus (CMV), infects fibroblast and endothelial cells and may cause mental retardation. CMV infections are more serious in immunocompromised patients. During infection, glycoprotein B (gB) plays an important role for viral entry and infection [184]. Kumar *et al.* [188] reported a potential diagnostic and therapeutic DNA aptamer against gB.

3. POTENTIALLY THERAPEUTIC APTAMERS AGAINST VIRAL PROTEINS

Development of drugs and vaccines against viruses is challenged by some factors, such as high mutation rate of viruses, low specificity, side effects and unexpected immune response of the host. These problems lead to the search for new treatment tools, that would be more effective and less dangerous. Several studies [132,189,190] have proven that aptamers successfully inhibit the penetration of viruses into cells and inhibit the enzymes involved in their replication.

3.1. Aptamers Against Human Immunodeficiency Virus (HIV) Proteins

3.1.1. Reverse Transcriptase (RT)

Reverse transcription has two enzymatic activities. The first is DNA polymerase activity and the other is RNase H activity that cleaves RNA [191]. In a study [192], inhibition effect of the 37 NT SELEX aptamer on the reverse transcriptase activity of HIV HXB2 strain was investigated *in vitro*. The 37 NT SELEX aptamer showed a high affinity for reverse transcriptase with a K_d value of 0.66 nM, whereas the random aptamer showed a K_d of ~ 161 nM. The authors found that 2.5 nM of the 37 NT SELEX aptamer was a potent inhibitor with about 50% reduced reaction rate. Based on conserved signature motifs, including family 1 pseudoknots (F1Pk), family 2 pseudoknots (F2Pk), 6/5 asymmetric loop motif and nonpseudoknot UCAA motif, several structural families were described for anti-

HIV RNA aptamers. The RNA aptamers were shown to inhibit the reverse transcripitase activity and block HIV replication. However, some aptamers were found to inhibit the enzymatic activity only in a few viral strains, while others showed broad spectrum inhibition [193,194]. In another study [195], the aptamers successfully inhibited the endogenous reverse transcriptase activity and infection in all three cell types infected with HIV-1 strain HXB2 (subtype B). Tuerk, *et al.* [196] performed SELEX and targeted the heterodimer structure of HIV1-RT. The identified aptamers had a consensus sequence resulting in formation of RNA pseudoknot. In another study, some biochemical studies and chemical modifications were performed to analyze the interaction between HIV-1-RT and pseudoknot RNA aptamer. The aptamer was shown to bind its target with a K_d value as low as ~ 25 pM [197]. Schneider *et al.* [198] developed a DNA aptamer named RT1t49, which could bind the reverse transcriptase from a subtype B strain of HIV-1 with a K_d value of 1 nM. Another study revealed the broad inhibitory capacity of the aptamers for both DNA polymerase and RNase H activities in HIV-1, SIVcpz, and HIV-2 strains RT1t49 [199].

Post-SELEX modifications were performed to increase the affinity HIV-1 RT aptamers. The heterodimer HIV-1 RT consists of a 66-kDa polypeptide chain (p66) complexed with a 51-kDa polypeptide chain (p51). p66 contains a groove in which duplex nucleic acid substrates could be placed between the polymerase and RNase H active sites. It was shown that 2'-O-methyl modified aptamer binds to p66 fingers and subdomains with a 10-fold higher affinity [200]. In another study, Somasunderam *et al.* [201] selected thioaptamers with thiophosphate backbone and showed that the selected aptamers could inhibit RNase H activity and viral replication. The thioaptamer R12-2 was shown to bind to HIV-1 RT with a K_d value of 70 nM.

3.1.2. Integrase (IN)

Retroviral integrase catalyzes insertion of viral DNA into the host cell's DNA. This property makes integrase an attractive target for development of antiretroviral drugs. The 17-mer oligonucleotide (ON) named T30177 was shown to inhibit HIV-1 integrase activity at nanomolar concentrations. A variant of T30177 named T30695 could also bind to HIV-integrase and inhibit its enzymatic activity [202,203]. Because of the structural similarities between RNase H and integrase, the effect of RNase H inhibitors on integrase were investigated. The truncated DNA aptamers (ODNs 93del and 112del) originated from ODN 93 and 112 could inhibit HIV-1 integrase in the nanomolar range. These G-rich aptamers couls form G-quartet structures stabilized with K^+ ions [204]. As potential anti-HIV therapeutic agents, various G-quadruplex RNA and DNA aptamers were shown to inhibit HIV-1 *in vitro* [205,206]. Based on molecular docking studies, a structural model of the complex between 93del and the tetramer of HIV-1 integrase was proposed. This strategy could be useful for design of HIV-1 integrase inhibitors [207]. In another study [208], RNA aptamers IN1, IN2, and IN3 showed K_d values ranging between 145 and 239 nmol/L, and formed a similar stem-loop structure, which is critical for binding to integrase. Although IN1 was not capable of forming the G-quartet structure, it could directly interact with the viral DNA binding site of the integrase. Rose *et al.* [209] identified 2'-deoxy--'-fluoroarabino nucleic acid (FANA) aptamers that could bind to HIV-1 integrase with affinity values in the range of 50–100 pM.

3.1.3. Other proteins of HIV-1

HIV-1 Rev is an essential viral regulatory protein that plays significant role in nuclear export of intron-containing viral RNA. Aptamers were developed against the Rev protein and shown to inhibit HIV-1 replication by interfering with Rev-RRE (Rev response element), Rev-Rev, and Rev-host protein interactions [210] [211]. In another study [211], RNA aptamer (RBA-14) with a K_d value of 5.9 nM was shown to block Rev oligomerization through the interaction between its major groove and arginine-rich helix of Rev.

The exterior envelope glycoprotein of HIV, gp120, plays an important role in viral entry through the interactions with CD4 receptor on host cells [212]. 2'-fluoropyrimidine-containing RNA (2'-F-RNA) aptamers with an affinity of 100 nM could bind to the gp120 protein of HIV-1 strain HXB2 and block the interaction with CD4 [213]. The selected 2'-F-RNA aptamers to HIV-1$_{BaL}$ gp120 showed neutralizing activity for group M and group O HIV-1 [214]. In addition, an aptamer named UCLA1 was identified as a HIV-1 entry inhibitor. The studies showed that B40 aptamer and its shortened derivatives (B40t77 and UCLA1) showed antiviral activity [215]. In another study, a novel non-SELEX screening approach was used for selection of aptamers. First, the affinities of gp120 aptamers were analysed by using HEX and HDOCK tools. The identified aptamers were then experimentally tested. The results revealed that the selected aptamers were able to inhibit HIV-1 infection up to 80% without any cytotoxicity [216].

In another study [217], it was shown that the synthetically simplified monomolecular Hotoda derivatives retained a remarkable antiviral activity. The study highlighted that monomolecular Hotoda's aptamers containing inversion of polarity sites represent a successful alternative strategy, which combines ease of synthesis and antiviral activity. The derivative containing two lipophilic groups, HT353LGly, could inhibit viral entry into the host cell with anti-HIV-1 activity in he low nanomolar range. However, other derivatives sharing the same basic sequence and similar topology were found ineffective.

HIV-1 is Gag polyprotein is another important structural protein, which is synthesized in the cytoplasm of infected cells. It is the only protein necessary and sufficient for formation of non-infectious virus-like particles (VLPs). Gag is co-translationally modified by N-terminal myristate, which helps passing through the plasma membrane of the infected cell. Gag is cleaved by the viral protease into its component proteins, namely matrix protein (MA), capsid protein (CA), nucleocapsid protein (NC), late domain (p6) and two small, differential peptides, SP1 and SP2. The aptamers selected against Gag protein were shown to play an important role in inhibition of HIV replication [218, 219].

Selection and characterization of RNA aptamers for HIV-1 Gag protein was also attempted by Ramalingam *et al.* [220]. The selected aptamers were able to specifically recognize the purified MA or NC proteins *in vitro,* as well as bind to a Gag protein lacking p6 (DP6-Gag). The K_d values of aptamers against Gag protein were found to range between 80 and 200 nM. Inhibitory effect anti-Gag aptamers on HIV-1 replication was tested, and it was shown that 7 of our 11 aptamers successfully decreased the level of extracellular p24. Based on the results, it was concluded that anti-Gag aptamers showed antiviral activity by decreasing the viral genomic RNA levels in the host cell.

3.2. Aptamers Against Hepatitis C Virus (HCV) Proteins

3.2.1. Non-Structural Proteins (NS) of HCV

Non-structural proteins such as NS2, NS3, NS4A, NS4B, NS5A and NS5B are viral proteins required for RNA replication [221]. RNA-dependent RNA polymerase (NS5B) is targeted to inhibit the replication of viral genomes. For this aim, aptamers were selected against NS5B. As a result, 2 of the aptamers were shown to specifically inhibit the polymerase activity of HCV NS5B *in vitro*. The aptamer 27v exhibited a K_d value in the nanomolar range [222]. Further studies were also performed to analyse the RNA motifs responsible for binding to NS5B. The results revealed that a single base change on the CGGG motif, also present in the stem structure of the NS5B coding RNA (5BSL3.2) was responsible for the change in binding affinity to NS5B [223 - 225].

In another study, Lee *et al.* [226] identified two types of 29 nucleotide-long NS5B-specific RNA aptamers with 2' hydroxyl nucleotides or 2' fluoro pyrimidines. Cellular internalization and toxicity, as well as affinity and specificity properties of the aptamers were tested. In a further study, 2'-fluoro pyrimidine-modified RNA aptamer against NS5B was conjugated with cholesterol for *in vivo* use. The results showed that the cholesterol-conjugated aptamer could efficiently enter into the cell and inhibit HCV RNA replication *in vitro*. *In vivo* studies of the cholesterol-conjugated RNA aptamers were also performed in the liver tissues of mice [227].

NS5A is an another protein important for HCV replication. The selected NS5A-4 and NS5A-5 aptamers were shown to inhibit viral RNA replication by preventing the interaction of NS5A with the core protein [228].

Non-structural HCV protein 3 (NS3) has both protease and helicase activities, which are necessary for HCV proliferation. This multifunctional enzyme is considered as a good target for development of antiviral aptamers. RNA aptamer that could bind to NS3 with a K_d value of 650 nM was shown to inhibit the protease activity [229]. In another study, the NS3-specific RNA aptamers specifically binding to the NS3 protease active site were selected. RNA aptamers G9-I, -II and -III, with a common GA (A / U) UGGGAC sequence, were effectively inhibited NS3 protease acitivity *in vitro* [230]. In addition, a G9 aptamer expression system was constructed in cultured cells for *in vivo* use of these aptamers by combining the cis-acting genomic human hepatitis delta virus (HDV) ribozyme and the G9-II aptamer. To generate the chimeric HDV-G9-II aptamer, the G9-II aptamer was inserted into the IV stem of the genomic HDV ribozyme. When applied in HeLa cells, HDV-G9-II aptamer showed higher protease inhibitory activity compared to G9-II, as revealed by Western-blot analysis [231].

In addition to studies related with the protease domain of HCV NS3, helicase activity was also targeted by aptamers. RNA aptamers were selected and shown to bind to HCV helicase with a K_d of 990 pM. Moreover, the RNA aptamers could block the binding of RNA, the substrate of HCV helicase, and can competitively inhibit helicase activity with high efficiency compared to poly (U) and tRNA. Inhibitory effect of the selected RNA aptamers on intracellular RNA synthesis of HCV replicon were also tested in human liver cells [232].

DNA aptamers against NS2 protein were selected by Gao *et al.* [233]. The results revealed that the aptamer NS2-2 showed antiviral activity by disrupting NS2/NS5A protein interaction. Moreover, the selected NS2-3 and NS2-2 aptamers were shown to inhibit production of infectious viruses without showing cytotoxicity in cell culture.

3.2.2. Structural Proteins of HCV

HCV envelope glycoproteins E1 and E2 are required for binding of virus to host cells. Therefore, E1 and E2 proteins are targets for development of antiviral drugs. For this aim, the aptamer ZE2 that could bind to E2 with high affinity and prevent infection in human hepatocytes by blocking the binding of E2 to the CD81 receptor was selected. The aptamer ZE2 with a K_d of 1.05 ± 0.4 nM showed highest binding affinity among other selected aptamers [129]. Binding sites of the aptamer on E2 were investigated in a further study [234].

DNA aptamers capable of binding both E1 and E2 proteins were selected. The results showed that the 50% effective concentrations (EC_{50}) of the aptamers E1E2-2, E1E2-4, E1E2-5 and E1E2-6 in virus-infected cells were 86.12 nM, 146.69 nM, 84.63 nM and 62.37 nM, respectively. In this study, it was shown that the aptamers developed against E1E2 could inhibit HCV infection in infectious cell culture system but show no effect on HCV replication in a replicon cell line [235].

3.3. Aptamers Against Hepatitis B Virus (HBV) Proteins

Apt.No.28, one of the five aptamers developed against the core protein of HBV (HBc), was shown to inhibit the replication of HBV by binding to core protein dimers and inhibiting nucleocapsid assembly, as revealed by agarose gel electrophoresis, capillary transfer and blotting methods [236]. In addition to oligonucleotide aptamers, the effect of peptide aptamers on HBV replication, capsid formation and virion production were investigated [237,238]. RNA aptamer named HBs-A22 that could bind specifically to Hepatitis B virus surface antigen (HBsAg) with high affinity aptamer was identified. The results showed that aptamer could bind specifically to HBsAg-expressing hematoma cell line HepG2.2.15, but not to HepG2 cells [239].

3.4. Aptamers Against Severe Acute Respiratory Syndrome (SARS) Coronavirus Proteins

NTPase / Helicase (nsP10), one of the known proteins of SARS coronavirus, is an interesting target for antiviral drug studies. The nsP10 protein targeting RNA aptamers with an IC_{50} of 1.2 nM were shown to inhibit the helicase activity up to 85% when measured by the fluorescence resonance energy transfer (FRET) based-assay, but it showed little effect on the ATPase activity in the presence of the cofactor poly (rU) [164]. Shum *et al.* [240] selected DNA aptamers against SARS-CoV helicase. As revealed by circular dichroism (CD) spectroscopy and gel electrophoresis, the aptamer showed G-quaternary and non-G-quaternary secondary structures. All selected aptamers with different IC_{50} values showed inhibition effect on SARS-CoV helicase. Interestingly, the non-G- quadruplex aptamers NG1, NG3 and NG8 had IC_{50} values of 87.7, 120.8 and 91.0 nM, respectively. However, the G-quadruplex aptamers G5 and G8 showed little inhibitory effect.

Since selection process of new aptamers is a time-consuming process, some studies were focused on testing the previously developed aptamers against new targets [241]. This kind of attempts are particularly important for the emerging SARS-CoV-2 [190]. BC007 was originally developed for neutralization of pathogenic autoantibodies to G-protein coupled receptors [242]. Binding of the BC007 aptamer to some crucial proteins of SARS-CoV-2 was tested [243].

DNA aptamer was selected against CoV2-S protein through an automatic selection process. Kinetics of the interaction between CoV2-S protein and the selected aptamers were analyzed by surface plasmon resonance (SPR), which revealed that aptamers could bind to CoV2-S with K_d values ranging between 9 and 21 nM. It was observed that the developed aptamers SP5, SP6 and SP7 could bind to CoV2-S with high specificity. In addition, the findings showed that aptamers inhibited viral infection independently of RBD, and inhibition was possible despite binding of the virus to cells [244].

3.5. Aptamers Against Influenza Virus Proteins

Influenza virus causes significant morbidity and mortality all over the world. Influenza surface glycoproteins, hemagglutinin (HA) and neuraminidase (NA), play active roles in the replication process of influenza A. While influenza C and B viruses mainly cause asymptomatic or mild respiratory infections, influenza A virus is the main cause of annual influenza outbreaks in humans [245,246].

Blocking the binding of the influenza virus surface glycoprotein hemagglutinin (HA) to the host cell receptors containing sialic acid is the main strategy to develop antiviral agents. DNA aptamer specific to the HA protein of H1N1 influenza virus (A / Puerto Rico / 8/1934) showed a high binding affinity (K_d = 78 ± 1 nM) [247]. Cheng *et al.* [248] targeted the HA1 protein of H5N1 influenza virus.

The selected aptamer showed strong binding inhibitory effect on the virus. In another study, RNA aptamer was developed by targeting the glycosylated receptor binding site of the HA protein (gHA1). RNA aptamer HA12-16 was shown to display the highest affinity, as revealed by nitrocellulose filter binding and enzyme-linked immunosorbent assay (ELISA) methods . In addition, it was observed that HA12-16 inhibited viral infection in host cells while increasing cell viability [249].

As a different approach, Gopinath *et al.* [149] used the whole virus rather than the purified proteins to develop aptamers. The aptamer P30-10-16 was shown to bind specifically to the HA region of the target strain A / Panama / 2007/1999 (H3N2), but not to other human influenza viruses. The P30-10-16 aptamer with a K_d of 188 pM showed 15 times higher affinity than the commercially available monoclonal antibodies.

Amino acid residues at the N-terminal of the PA subunit (PA_N) of the influenza A virus polymerase are highly conserved among different subtypes of the influenza virus. This region plays a very important role in terms of endonuclease activity, protein stability and viral RNA (vRNA) promoter binding. This makes PA_N an attractive target for developing anti-influenza agents. Yuan *et al.* [250] performed SELEX by targeting the PA protein or PA_N domain of H5N1. As a result, 3 aptamers against PA and 6 aptamers against PAN were identified. The aptamers selected for PA did not show neither endonuclease inhibitory effect nor antiviral effects. In contrast, 4 of the six aptamers (PAN-1, PAN-2, PAN-3, and PAN-4) developed against PAN were shown to inhibit both endonuclease activity and H5N1 virus infection. Moreover, the PAN-2 aptamer with a K_d of 247 ± 11 nM also showed cross-protection against the H1N1, H5N1, H7N7 and H7N9 viruses.

4. CONCLUDING REMARKS

Covid-19 pandemic highlighted once more that we need fast and sensitive detection methods, as well as effective drugs targeting novel virulence mechanisms and virulence proteins. Here, nucleic acid and peptide aptamers offer great potential as they can be developed against specific protein targets by using *in vitro* methods. Aptamers with comparable binding affinity and specificity are strong alternatives to antibodies. Compared to the antibody production process, *in vitro* SELEX and display methods eliminate the need for using live cells, thereby decreasing the time and cost of the production process. Moreover, due to their small size, aptamers can be chemically synthesized and modified with functional groups. In addition, small size reduces the risk of undesired immune response unlike the case of antibodies. In conclusion, the drug potential of aptamers may offer a solution for immediate response to pandemics in the future. On the other hand, the weaknesses of aptamers should not be unnoticed. Nucleic acid aptamers and peptide aptamers are prone to degradation in serum by nucleases and proteases, respectively. Another limitation is renal filtration of aptamers due to their small size. Therefore, improvement and a careful evaluation of pharmacokinetic properties are important as much as the aptamer selection process. In order to overcome those limitations, chemical modification of aptamers, as well as conjugation with appropriate partners to increase serum half-life and impart penetration capability across cellular barriers, could be some alternative strategies. Although aptamers are promising agents, small-molecule drugs still dominate the antiviral drug market. Therefore, more studies are needed to prove the *in vivo* efficiency of aptamers.

CONSENT FOR PUBLICATION

Not applicable.

CONFLICT OF INTEREST

The author declares no conflict of interest, financial or otherwise.

ACKNOWLEDGEMENTS

Declared none.

REFERENCES

[1]   Jijakli K, Khraiwesh B, Fu W, Luo L, Alzahmi A, Koussa J, *et al.* The *in vitro* selection world. Methods 2016; 106: 3-13.

[2]   Amstutz P, Forrer P, Zahnd C, Plückthun A. *In vitro* display technologies: Novel developments and applications. Curr Opin Biotechnol 2001; 12: 400-5.

[3]   Dotter H, Boll M, Eder M, Eder A-C. Library and post-translational modifications of peptide-based display systems. Biotechnol Adv 2021; 47: 107699.
[http://dx.doi.org/10.1016/j.biotechadv.2021.107699] [PMID: 33513435]

[4]   Ohashi H, Shimizu Y, Ying BW, Ueda T. Efficient protein selection based on ribosome display system with purified components. Biochem Biophys Res Commun 2007; 352(1): 270-6.
[http://dx.doi.org/10.1016/j.bbrc.2006.11.017] [PMID: 17113037]

[5]   Wang W, Hara S, Liu M, Aigaki T, Shimizu S, Ito Y. Polypeptide aptamer selection using a stabilized ribosome display. J Biosci Bioeng. 2011 Nov 1;112(5):515–7. using mRNA-display. Methods 2013; 60(1): 55-69.
[PMID: 23201412]

[6]   Bashiruddin NK, Suga H. Construction and screening of vast libraries of natural product-like macrocyclic peptides using *in vitro* display technologies. Curr Opin Chem Biol 2015; 24: 131-8.

[7]   Chin SE, Ferraro F, Groves M, Liang M, Vaughan TJ, Dobson CL. Isolation of high-affinity, neutralizing anti-idiotype antibodies by phage and ribosome display for application in immunogenicity and pharmacokinetic analyses. J Immunol Methods 2015; 416: 49-58.
[http://dx.doi.org/10.1016/j.jim.2014.10.013] [PMID: 25449532]

[8]   Valencia CA, Zou J, Liu R. *In vitro* selection of proteins with desired characteristics using mRNA-display. Methods 2013; 60(1): 55-69.
[http://dx.doi.org/10.1016/j.ymeth.2012.11.004] [PMID: 23201412]

[9]   Dower WJ, Mattheakis LC. *In vitro* selection as a powerful tool for the applied evolution of proteins and peptides. Curr Opin Chem Biol 2002; 6: 390-8.

[http://dx.doi.org/10.1016/S1367-5931(02)00332-0]

[10] Yang KA, Pei R, Stojanovic MN. *In vitro* selection and amplification protocols for isolation of aptameric sensors for small molecules. Methods 2016; 106: 58-65.
[http://dx.doi.org/10.1016/j.ymeth.2016.04.032]

[11] Pobanz K, Lupták A. Improving the odds: Influence of starting pools on *in vitro* selection outcomes. Methods 2016; 106: 14-20.

[12] Van Dorst B, Mehta J, Rouah-Martin E, Blust R, Robbens J. Phage display as a method for discovering cellular targets of small molecules. Methods 2012; 58: 56-61.
[http://dx.doi.org/10.1016/j.ymeth.2012.07.011]

[13] Ihara H, Mie M, Funabashi H, *et al.* In vitro selection of zinc finger DNA-binding proteins through ribosome display. Biochem Biophys Res Commun 2006; 345(3): 1149-54.
[http://dx.doi.org/10.1016/j.bbrc.2006.05.029] [PMID: 16714002]

[14] Josephson K, Ricardo A, Szostak JW. MRNA display: From basic principles to macrocycle drug discovery. Drug Discovery Today 2014; 19: 388-99.

[15] Blanco C, Verbanic S, Seelig B, Chen IA. High throughput sequencing of *in vitro* selections of mRNA-displayed peptides: data analysis and applications. Phys Chem Chem Phys 2020; 22(12): 6492-506.
[http://dx.doi.org/10.1039/C9CP05912A] [PMID: 31967131]

[16] Bayat P, Nosrati R, Alibolandi M, Rafatpanah H, Abnous K, Khedri M, *et al.* SELEX methods on the road to protein targeting with nucleic acid aptamers. Biochimie 2018; Vol. 154: 132-55.

[17] Wada A. Ribosome display technology for selecting peptide and protein ligands. Biomedical Applications of Functionalized Nanomaterials: Concepts, Development and Clinical Translation 2018; 89-104.
[http://dx.doi.org/10.1016/B978-0-323-50878-0.00004-5]

[18] Huang Y, Wiedmann MM, Suga H. RNA display methods for the discovery of bioactive macrocycles. Chem Rev 2019; 119(17): 10360-91.
[http://dx.doi.org/10.1021/acs.chemrev.8b00430] [PMID: 30395448]

[19] Galán A, Comor L, Horvatić A, *et al.* Library-based display technologies: where do we stand? Mol Biosyst 2016; 12(8): 2342-58.
[http://dx.doi.org/10.1039/C6MB00219F] [PMID: 27306919]

[20] Nagumo Y, Fujiwara K, Horisawa K, Yanagawa H, Doi N. PURE mRNA display for *in vitro* selection of single-chain antibodies. J Biochem 2016; 159(5): 519-26.
[http://dx.doi.org/10.1093/jb/mvv131] [PMID: 26711234]

[21] Shimizu Y, Inoue A, Tomari Y, *et al.* Cell-free translation reconstituted with purified components. Nat Biotechnol 2001; 19(8): 751-5.
[http://dx.doi.org/10.1038/90802] [PMID: 11479568]

[22] Newton MS, Cabezas-Perusse Y, Tong CL, Seelig B. *In vitro* selection of peptides and proteins-advantages of mrna display. ACS Synth Biol 2020; 9(2): 181-90.
[http://dx.doi.org/10.1021/acssynbio.9b00419] [PMID: 31891492]

[23] Sci TB. Chapter 2 mRNA Display : Ligand Discovery. Interaction Analysis 2003; 159-65.

[24] Seelig B. mRNA display for the selection and evolution of enzymes from *in vitro*-translated protein libraries. Nat Protoc 2011; 6(4): 540-52.
[http://dx.doi.org/10.1038/nprot.2011.312] [PMID: 21455189]

[25] Lipovsek D, Plückthun A. *In-vitro* protein evolution by ribosome display and mRNA display. J Immunol Methods 2004; 290(1-2): 51-67.
[http://dx.doi.org/10.1016/j.jim.2004.04.008] [PMID: 15261571]

[26] Takahashi TT, Austin RJ, Roberts RW. mRNA display: ligand discovery, interaction analysis and

beyond. Trends Biochem Sci 2003; 28(3): 159-65.
[http://dx.doi.org/10.1016/S0968-0004(03)00036-7] [PMID: 12633996]

[27]   Wang H, Liu R. Advantages of mRNA display selections over other selection techniques for investigation of protein-protein interactions. Expert Rev Proteomics 2011; 8(3): 335-46.
[http://dx.doi.org/10.1586/epr.11.15] [PMID: 21679115]

[28]   Oikonomou P, Salatino R, Tavazoie S. *In vivo* mRNA display enables large-scale proteomics by next generation sequencing. Proc Natl Acad Sci USA 2020; 117(43): 26710-8.
[http://dx.doi.org/10.1073/pnas.2002650117] [PMID: 33037152]

[29]   Mattheakis LC, Bhatt RR, Dower WJ. An *in vitro* polysome display system for identifying ligands from very large peptide libraries. Proc Natl Acad Sci USA 1994; 91(19): 9022-6.
[http://dx.doi.org/10.1073/pnas.91.19.9022] [PMID: 7522328]

[30]   Li R, Kang G, Hu M, Huang H. Ribosome display: a potent display technology used for selecting and evolving specific binders with desired properties. Mol Biotechnol 2019; 61(1): 60-71.
[http://dx.doi.org/10.1007/s12033-018-0133-0] [PMID: 30406440]

[31]   He M, Edwards BM, Kastelic D, Taussig MJ. Ribosome display and related technologies. Methods and Protocols 2012; 805: 75-85.
[http://dx.doi.org/10.1007/978-1-61779-379-0_5]

[32]   He M, Taussig MJ. Ribosome display: cell-free protein display technology. Brief Funct Genomics Proteomics 2002; 1(2): 204-12.
[http://dx.doi.org/10.1093/bfgp/1.2.204] [PMID: 15239905]

[33]   Kanamori T, Fujino Y, Ueda T. PURE ribosome display and its application in antibody technology. Biochim Biophys Acta 2014; 1844(11): 1925-32.
[http://dx.doi.org/10.1016/j.bbapap.2014.04.007] [PMID: 24747149]

[34]   Zhao F, Shi R, Liu R, Tian Y, Yang Z. Application of phage-display developed antibody and antigen substitutes in immunoassays for small molecule contaminants analysis: A mini-review. Food Chemistry 2021; 339: 128084.

[35]   Yan X, Xu Z. Ribosome-display technology: applications for directed evolution of functional proteins. Drug Discov Today 2006; 11(19-20): 911-6.
[http://dx.doi.org/10.1016/j.drudis.2006.08.012] [PMID: 16997141]

[36]   Zhao XL, Chen WQ, Yang ZH, Li JM, Zhang SJ, Tian LF. Selection and affinity maturation of human antibodies against rabies virus from a scFv gene library using ribosome display. J Biotechnol 2009; 144(4): 253-8.
[http://dx.doi.org/10.1016/j.jbiotec.2009.09.022] [PMID: 19818816]

[37]   Luo Y, Xia Y. Selection of single-chain variable fragment antibodies against fenitrothion by ribosome display. Anal Biochem 2012; 421(1): 130-7.
[http://dx.doi.org/10.1016/j.ab.2011.10.044] [PMID: 22138186]

[38]   Kunamneni A, Ye C, Bradfute SB, Durvasula R. Ribosome display for the rapid generation of high-affinity Zika-neutralizing single-chain antibodies. PLoS One 2018; 13(11): e0205743.
[http://dx.doi.org/10.1371/journal.pone.0205743] [PMID: 30444865]

[39]   Leipert TK, Baldeschwieler JD, Shirley DA. Applications of gamma ray angular correlations to the study of biological macromolecules in solution. Nature 1968; 220(5170): 907-9.
[http://dx.doi.org/10.1038/220907a0] [PMID: 5722139]

[40]   Zahnd C, Amstutz P, Plückthun A. Ribosome display: selecting and evolving proteins *in vitro* that specifically bind to a target. Nat Methods 2007; 4(3): 269-79.
[http://dx.doi.org/10.1038/nmeth1003] [PMID: 17327848]

[41]   Irving RA, Coia G, Roberts A, Nuttall SD, Hudson PJ. Ribosome display and affinity maturation: from antibodies to single V-domains and steps towards cancer therapeutics. J Immunol Methods 2001; 248(1-2): 31-45.

[http://dx.doi.org/10.1016/S0022-1759(00)00341-0] [PMID: 11223067]

[42]   Reverdatto S, Burz DS, Shekhtman A. Peptide aptamers: development and applications. Curr Top Med Chem 2015; 15(12): 1082-101.
[http://dx.doi.org/10.2174/1568026615666150413153143] [PMID: 25866267]

[43]   Baines IC, Colas P. Peptide aptamers as guides for small-molecule drug discovery. Drug Discov Today 2006; 11(7-8): 334-41.
[http://dx.doi.org/10.1016/j.drudis.2006.02.007] [PMID: 16580975]

[44]   Hoppe-Seyler F, Crnkovic-Mertens I, Tomai E, Butz K. Peptide aptamers: specific inhibitors of protein function. Curr Mol Med 2004; 4(5): 529-38.
[http://dx.doi.org/10.2174/1566524043360519] [PMID: 15267224]

[45]   Hoppe-Seyler F, Butz K. Peptide aptamers: powerful new tools for molecular medicine. J Mol Med (Berl) 2000; 78(8): 426-30.
[http://dx.doi.org/10.1007/s001090000140] [PMID: 11097111]

[46]   Li J, Tan S, Chen X, Zhang C-Y, Zhang Y. Peptide aptamers with biological and therapeutic applications. Curr Med Chem 2011; 18(27): 4215-22.
[http://dx.doi.org/10.2174/092986711797189583] [PMID: 21838684]

[47]   Blind M, Blank M. Aptamer selection technology and recent advances. Mol Ther Nucleic Acids 2015; 4(1): e223.
[http://dx.doi.org/10.1038/mtna.2014.74] [PMID: 28110747]

[48]   Wang ZJ, Chen EN, Yang G, Zhao XY, QU F. Research advances of aptamers selection for small molecule targets. Chin J Anal Chem 2020; 48(5): 573-82.
[http://dx.doi.org/10.1016/S1872-2040(20)60013-5]

[49]   Ellington AD, Szostak JW. *In vitro* selection of RNA molecules that bind specific ligands. Nature 1990; 346(6287): 818-22.
[http://dx.doi.org/10.1038/346818a0] [PMID: 1697402]

[50]   Tuerk C, Gold L. Systematic evolution of ligands by exponential enrichment: RNA ligands to bacteriophage T4 DNA polymerase. Science (80-) 1990; 249(4968): 505-10.

[51]   Song KM, Lee S, Ban C. Aptamers and their biological applications. Sensors (Basel) 2012; 12(1): 612-31.
[http://dx.doi.org/10.3390/s120100612] [PMID: 22368488]

[52]   Lakhin AV, Tarantul VZ, Gening LV. Aptamers: problems, solutions and prospects. Acta Naturae 2013; 5(4): 34-43.
[http://dx.doi.org/10.32607/20758251-2013-5-4-34-43] [PMID: 24455181]

[53]   Odeh F, Nsairat H, Alshaer W, *et al.* Aptamers chemistry: chemical modifications and conjugation strategies. Molecules 2019; 25(1): 1-51.
[http://dx.doi.org/10.3390/molecules25010003] [PMID: 31861277]

[54]   Sun H, Zu Y. A Highlight of recent advances in aptamer technology and its application. Molecules 2015; 20(7): 11959-80.
[http://dx.doi.org/10.3390/molecules200711959] [PMID: 26133761]

[55]   Zon G. Recent advances in aptamer applications for analytical biochemistry. Anal Biochem 2020; (June): 113894.
[http://dx.doi.org/10.1016/j.ab.2020.113894] [PMID: 32763306]

[56]   Szeitner Z, András J, Gyurcsányi RE, Mészáros T. Is less more? Lessons from aptamer selection strategies. J Pharm Biomed Anal 2014; 101: 58-65.
[http://dx.doi.org/10.1016/j.jpba.2014.04.018] [PMID: 24877649]

[57]   Zhang Y, Lai BS, Juhas M. Recent advances in aptamer discovery and applications. Molecules 2019; 24(5): E941.

[http://dx.doi.org/10.3390/molecules24050941] [PMID: 30866536]

[58]   Mascini M, Palchetti I, Tombelli S. Nucleic acid and peptide aptamers: fundamentals and bioanalytical aspects. Angew Chem Int Ed Engl 2012; 51(6): 1316-32.
[http://dx.doi.org/10.1002/anie.201006630] [PMID: 22213382]

[59]   Zhou J, Rossi J. Aptamers as targeted therapeutics: current potential and challenges have a patent pending on "Cell-specific internalizing RNA aptamers against human CCR5 and used therefore" [United States Patent HHS Public Access. Nat Rev Drug Discov 2017; 16(3): 181-202.

[60]   Meng HM, Liu H, Kuai H, Peng R, Mo L, Zhang XB. Aptamer-integrated DNA nanostructures for biosensing, bioimaging and cancer therapy. Chem Soc Rev 2016; 45(9): 2583-602.
[http://dx.doi.org/10.1039/C5CS00645G] [PMID: 26954935]

[61]   Bala J, Chinnapaiyan S, Dutta RK, Unwalla H. Aptamers in HIV research diagnosis and therapy. RNA Biol 2018; 15(3): 327-37.
[http://dx.doi.org/10.1080/15476286.2017.1414131] [PMID: 29431588]

[62]   Ning Y, Hu J, Lu F. Since January 2020 Elsevier has created a COVID-19 resource centre with free information in English and Mandarin on the novel coronavirus COVID- 19 . The COVID-19 resource centre is hosted on Elsevier Connect , the company ' s public news and information 2020.

[63]   Ismail SI, Alshaer W. Therapeutic aptamers in discovery, preclinical and clinical stages. Adv Drug Deliv Rev 2018; 134: 51-64.
[http://dx.doi.org/10.1016/j.addr.2018.08.006] [PMID: 30125605]

[64]   Bouvier-Müller A, Ducongé F. Application of aptamers for *in vivo* molecular imaging and theranostics. Adv Drug Deliv Rev 2018; 134: 94-106.
[http://dx.doi.org/10.1016/j.addr.2018.08.004] [PMID: 30125606]

[65]   Zou X, Wu J, Gu J, Shen L, Mao L. Application of aptamers in virus detection and antiviral therapy. Front Microbiol 2019; 10: 1462.
[http://dx.doi.org/10.3389/fmicb.2019.01462] [PMID: 31333603]

[66]   Pan Q, Luo F, Liu M, Zhang XL. Oligonucleotide aptamers: promising and powerful diagnostic and therapeutic tools for infectious diseases. J Infect 2018; 77(2): 83-98.
[http://dx.doi.org/10.1016/j.jinf.2018.04.007] [PMID: 29746951]

[67]   Kanwar JR, Roy K, Kanwar RK. Chimeric aptamers in cancer cell-targeted drug delivery. Crit Rev Biochem Mol Biol 2011; 46(6): 459-77.
[http://dx.doi.org/10.3109/10409238.2011.614592] [PMID: 21955150]

[68]   Perret G, Boschetti E. Aptamer affinity ligands in protein chromatography. Biochimie 2018; 145: 98-112.
[http://dx.doi.org/10.1016/j.biochi.2017.10.008] [PMID: 29054800]

[69]   Groff K, Brown J, Clippinger AJ. Modern affinity reagents: Recombinant antibodies and aptamers. Biotechnol Adv 2015; 33(8): 1787-98.
[http://dx.doi.org/10.1016/j.biotechadv.2015.10.004] [PMID: 26482034]

[70]   Bhardwaj R, Krautz-Peterson G, Skelly PJ. Using RNA interference in *Schistosoma mansoni*. Methods Mol Biol 2011; 764(3): 223-39.
[http://dx.doi.org/10.1007/978-1-61779-188-8_15] [PMID: 21748644]

[71]   Ning Y, Hu J, Lu F. Aptamers used for biosensors and targeted therapy. Biomed Pharmacother 2020; 132(August): 110902.
[http://dx.doi.org/10.1016/j.biopha.2020.110902] [PMID: 33096353]

[72]   Darmostuk M, Rimpelova S, Gbelcova H, Ruml T. Current approaches in SELEX: An update to aptamer selection technology. Biotechnol Adv 2015; 33(6 Pt 2): 1141-61.
[http://dx.doi.org/10.1016/j.biotechadv.2015.02.008] [PMID: 25708387]

[73]   Kaur H. Recent developments in cell-SELEX technology for aptamer selection. Biochim Biophys

Acta, Gen Subj 2018; 1862(10): 2323-9.
[http://dx.doi.org/10.1016/j.bbagen.2018.07.029] [PMID: 30059712]

[74]   Amraee M, Oloomi M, Yavari A, Bouzari S. DNA aptamer identification and characterization for E. coli O157 detection using cell based SELEX method. Anal Biochem 2017; 536: 36-44.
[http://dx.doi.org/10.1016/j.ab.2017.08.005] [PMID: 28818557]

[75]   He J, Wang J, Zhang N, *et al. In vitro* selection of DNA aptamers recognizing drug-resistant ovarian cancer by cell-SELEX. Talanta 2019; 194: 437-45.
[http://dx.doi.org/10.1016/j.talanta.2018.10.028] [PMID: 30609555]

[76]   Sefah K, Shangguan D, Xiong X, O'Donoghue MB, Tan W. Development of DNA aptamers using cell-SELEX. Nat Protoc 2010; 5(6): 1169-85.
[http://dx.doi.org/10.1038/nprot.2010.66] [PMID: 20539292]

[77]   Bing T, Zhang N, Shangguan D. Cell-SELEX, an effective way to the discovery of biomarkers and unexpected molecular events. Adv Biosyst 2019; 3(12): e1900193.
[http://dx.doi.org/10.1002/adbi.201900193] [PMID: 32648677]

[78]   Li W, Wang S, Zhou L, Cheng Y, Fang J. An ssDNA aptamer selected by Cell-SELEX for the targeted imaging of poorly differentiated gastric cancer tissue. Talanta 2019; 199: 634-42.
[http://dx.doi.org/10.1016/j.talanta.2019.03.016] [PMID: 30952308]

[79]   Ohuchi S. Cell-SELEX Technology. Biores Open Access 2012; 1(6): 265-72.
[http://dx.doi.org/10.1089/biores.2012.0253] [PMID: 23515081]

[80]   Sola M, Menon AP, Moreno B, *et al.* Aptamers against live targets: is *in vivo* SELEX finally coming to the edge? Mol Ther Nucleic Acids 2020; 21: 192-204.
[http://dx.doi.org/10.1016/j.omtn.2020.05.025] [PMID: 32585627]

[81]   Pleiko K, Saulite L, Parfejevs V, Miculis K, Vjaters E, Riekstina U. Differential binding cell-SELEX method to identify cell-specific aptamers using high-throughput sequencing. Sci Rep 2019; 9(1): 8142.
[http://dx.doi.org/10.1038/s41598-019-44654-w] [PMID: 31148584]

[82]   Catuogno S, Esposito CL. Aptamer cell-based selection: Overview and advances. Biomedicines 2017; 5(3): E49.
[http://dx.doi.org/10.3390/biomedicines5030049] [PMID: 28805744]

[83]   Bruno JG, Kiel JL. Use of magnetic beads in selection and detection of biotoxin aptamers by electrochemiluminescence and enzymatic methods. Biotechniques 2002; 32(1): 178-180, 182-183.
[http://dx.doi.org/10.2144/02321dd04] [PMID: 11808691]

[84]   Szeto K, Craighead HG. Devices and approaches for generating specific high-affinity nucleic acid aptamers. Appl Phys Rev 2014; 1(3)
[http://dx.doi.org/10.1063/1.4894851]

[85]   Wang M, Wang Q, Li X, Lu L, Du S, Zhang H. Selection and identification of diethylstilbestrol-specific aptamers based on magnetic-bead SELEX. Microchem J
[http://dx.doi.org/10.1016/j.microc.2020.105354]

[86]   Qiao L, Wang H, He J, Yang S, Chen A. Truncated affinity-improved aptamers for 17β-estradiol determination by AuNPs-based colorimetric aptasensor. Food Chem 2021; 340: 128181.
[http://dx.doi.org/10.1016/j.foodchem.2020.128181] [PMID: 33032145]

[87]   Ge MH, Zhu XH, Shao YM, *et al.* Synthesis and characterization of CD133 targeted aptamer-drug conjugates for precision therapy of anaplastic thyroid cancer. Biomater Sci 2020.
[PMID: 33350399]

[88]   Esposito CL, Quintavalle C, Ingenito F, *et al.* Identification of a novel RNA aptamer that selectively targets breast cancer exosomes. Mol Ther Nucleic Acids 2021; 23: 982-94.
[http://dx.doi.org/10.1016/j.omtn.2021.01.012] [PMID: 33614245]

[89]   Farhana Lipi, Suxiang Chen, Madhuri Chakravarthy, Shilpa Rakesh, Rakesh N Veedu. *In vitro*

evolution of chemically-modified nucleic acid aptamers: Pros and cons, and comprehensive selection strategies. RNA Biol 2016; 13(12): 1232-45.

[90]   Tolnai ZJ, András J, Szeitner Z, *et al.* Spiegelmer-based sandwich assay for cardiac troponin I detection. Int J Mol Sci 2020; 21(14): 4963.
[http://dx.doi.org/10.3390/ijms21144963] [PMID: 32674303]

[91]   Williams KP, Liu XH, Schumacher TN, *et al.* Bioactive and nuclease-resistant L-DNA ligand of vasopressin. Proc Natl Acad Sci USA 1997; 94(21): 11285-90.
[http://dx.doi.org/10.1073/pnas.94.21.11285] [PMID: 9326601]

[92]   Yan X, Gao X, Zhang Z. Isolation and characterization of 2'-amino-modified RNA aptamers for human TNFalpha. Genomics Proteomics Bioinformatics 2004; 2(1): 32-42.
[http://dx.doi.org/10.1016/S1672-0229(04)02005-4] [PMID: 15629041]

[93]   Lin Y, Nieuwlandt D, Magallanez A, Feistner B, Jayasena SD. High-affinity and specific recognition of human thyroid stimulating hormone (hTSH) by *in vitro*-selected 2'-amino-modified RNA. Nucleic Acids Res 1996; 24(17): 3407-14.
[http://dx.doi.org/10.1093/nar/24.17.3407] [PMID: 8811096]

[94]   Thirunavukarasu D, Chen T, Liu Z, Hongdilokkul N, Romesberg FE. Selection of 2'-fluoro-modified aptamers with optimized properties. J Am Chem Soc 2017; 139(8): 2892-5.
[http://dx.doi.org/10.1021/jacs.6b13132] [PMID: 28218835]

[95]   Hervas-Stubbs S, Soldevilla MM, Villanueva H, Mancheño U, Bendandi M, Pastor F. Identification of TIM3 2'-fluoro oligonucleotide aptamer by HT-SELEX for cancer immunotherapy. Oncotarget 2016; 7(4): 4522-30.
[http://dx.doi.org/10.18632/oncotarget.6608] [PMID: 26683225]

[96]   Lee J-H, Canny MD, De Erkenez A, *et al.* A therapeutic aptamer inhibits angiogenesis by specifically targeting the heparin binding domain of VEGF165. Proc Natl Acad Sci USA 2005; 102(52): 18902-7.
[http://dx.doi.org/10.1073/pnas.0509069102] [PMID: 16357200]

[97]   Tucker CE, Chen L-S, Judkins MB, Farmer JA, Gill SC, Drolet DW. Detection and plasma pharmacokinetics of an anti-vascular endothelial growth factor oligonucleotide-aptamer (NX1838) in rhesus monkeys. J Chromatogr B Biomed Sci Appl 1999; 732(1): 203-12.
[http://dx.doi.org/10.1016/S0378-4347(99)00285-6] [PMID: 10517237]

[98]   Lebars I, Richard T, Di Primo C, Toulmé JJ. LNA derivatives of a kissing aptamer targeted to the trans-activating responsive RNA element of HIV-1. Blood Cells Mol Dis 2007; 38(3): 204-9.
[http://dx.doi.org/10.1016/j.bcmd.2006.11.008] [PMID: 17300966]

[99]   Nonaka Y, Yoshida W, Abe K, *et al.* Affinity improvement of a VEGF aptamer by *in silico* maturation for a sensitive VEGF-detection system. Anal Chem 2013; 85(2): 1132-7.
[http://dx.doi.org/10.1021/ac303023d] [PMID: 23237717]

[100]   Reinemann C, Strehlitz B. Aptamer-modified nanoparticles and their use in cancer diagnostics and treatment. Swiss Med Wkly 2014; 144w13908.
[http://dx.doi.org/10.4414/smw.2014.13908]

[101]   Antisense drug discovery and development. UK: Future Science Ltd London 2021.

[102]   R E Wang, H Wu, Y Niu, J Cai. Improving the stability of aptamers by chemical modification. Bentham Science Publishers 2011.

[103]   Saccà B, Lacroix L, Mergny JL. The effect of chemical modifications on the thermal stability of different G-quadruplex-forming oligonucleotides. Nucleic Acids Res 2005; 33(4): 1182-92.
[http://dx.doi.org/10.1093/nar/gki257] [PMID: 15731338]

[104]   Abeydeera ND, Egli M, Cox N, *et al.* Evoking picomolar binding in RNA by a single phosphorodithioate linkage. Nucleic Acids Res 2016; 44(17): 8052-64.
[http://dx.doi.org/10.1093/nar/gkw725] [PMID: 27566147]

[105]   Gold L, Ayers D, Bertino J, *et al.* Aptamer-based multiplexed proteomic technology for biomarker discovery. Nat Prec 2010; p. 1.

[106]   Luo Y, Wadhawan S, Greenfield A, *et al.* SOMAscan proteomics identifies serum biomarkers associated with liver fibrosis in patients with NASH. Hepatol Commun 2021.

[107]   Katilius E, Carmel AB, Koss H, *et al.* Sperm cell purification from mock forensic swabs using SOMAmer™ affinity reagents. Forensic Sci Int Genet 2018; 35: 9-13.
[http://dx.doi.org/10.1016/j.fsigen.2018.03.011] [PMID: 29609058]

[108]   Franklin MC, Kadkhodayan S, Ackerly H, *et al.* Structure and function analysis of peptide antagonists of melanoma inhibitor of apoptosis (ML-IAP). Biochemistry 2003; 42(27): 8223-31.
[http://dx.doi.org/10.1021/bi034227t] [PMID: 12846571]

[109]   Chorev M, Goodman M. A dozen years of retro-inverso peptidomimetics. Acc Chem Res 1993; 26(5): 266-73.
[http://dx.doi.org/10.1021/ar00029a007]

[110]   Li P, Roller P. Cyclization strategies in peptide derived drug design. Bentham Science Publishers 2002.
[http://dx.doi.org/10.2174/1568026023394209]

[111]   Taylor EM, Otero DA, Banks WA, O'Brien JS. Retro-inverso prosaptide peptides retain bioactivity, are stable *In vivo*, and are blood-brain barrier permeable. J Pharmacol Exp Ther 2000; 295(1): 190-4.
[PMID: 10991978]

[112]   Wang RE, Wu H, Niu Y, Cai J. Improving the stability of aptamers by chemical modification. Curr Med Chem 2011; 18(27): 4126-38.
[http://dx.doi.org/10.2174/092986711797189565] [PMID: 21838692]

[113]   Krammer F, Smith GJD, Fouchier RAM, *et al.* Influenza. Nat Rev Dis Primers 2018; 4(1): 3.
[http://dx.doi.org/10.1038/s41572-018-0002-y] [PMID: 29955068]

[114]   Udugama B, Kadhiresan P, Kozlowski HN, *et al.* Diagnosing COVID-19: the disease and tools for detection. ACS Nano 2020; 14(4): 3822-35.
[http://dx.doi.org/10.1021/acsnano.0c02624] [PMID: 32223179]

[115]   Zhang W, Du RH, Li B, *et al.* Molecular and serological investigation of 2019-nCoV infected patients: implication of multiple shedding routes. Emerg Microbes Infect 2020; 9(1): 386-9.
[http://dx.doi.org/10.1080/22221751.2020.1729071] [PMID: 32065057]

[116]   Pérez-Ruiz M, Pedrosa-Corral I, Sanbonmatsu-Gámez S, Navarro-Marí M. Laboratory detection of respiratory viruses by automated techniques. Open Virol J 2012; 6(1): 151-9.
[http://dx.doi.org/10.2174/1874357901206010151] [PMID: 23248735]

[117]   Zheng H, Lang Y, Yu J, Han Z, Chen B, Wang Y. Affinity binding of aptamers to agarose with DNA tetrahedron for removal of hepatitis B virus surface antigen. Colloids Surf B Biointerfaces 2019; 178(178): 80-6. [Internet].
[http://dx.doi.org/10.1016/j.colsurfb.2019.02.040] [PMID: 30844563]

[118]   Xi Z, Huang R, Li Z, *et al.* Selection of HBsAg-specific DNA aptamers based on carboxylated magnetic nanoparticles and their application in the rapid and simple detection of hepatitis b virus infection. ACS Appl Mater Interfaces 2015; 7(21): 11215-23.
[http://dx.doi.org/10.1021/acsami.5b01180] [PMID: 25970703]

[119]   Michler T, Kosinska AD, Festag J, *et al.* Knockdown of virus antigen expression increases therapeutic vaccine efficacy in high-titer hepatitis b virus carrier mice. Gastroenterology 2020; 158(6): 1762-1775.e9. [Internet].
[http://dx.doi.org/10.1053/j.gastro.2020.01.032] [PMID: 32001321]

[120]   Mohsin DH, Mashkour MS, Fatemi F. Design of aptamer-based sensing platform using gold nanoparticles functionalized reduced graphene oxide for ultrasensitive detection of Hepatitis B virus.

Chem Pap 2020; 0123456789.

[121]   Liu Y, Le C, Tyrrell DL, Le XC, Li XF. Aptamer binding assay for the e antigen of hepatitis b using modified aptamers with g-quadruplex structures. Anal Chem 2020; 92(9): 6495-501.
[http://dx.doi.org/10.1021/acs.analchem.9b05740] [PMID: 32250595]

[122]   Huang R, Xi Z, Deng Y, He N. Fluorescence based Aptasensors for the determination of hepatitis B virus e antigen. Sci Rep 2016; 6(August): 31103.
[http://dx.doi.org/10.1038/srep31103] [PMID: 27499342]

[123]   Malsagova KA, Pleshakova TO, Galiullin RA, Shumov ID, Kozlo AF, Romanova TS, *et al.* Nanowire aptamer-sensitized biosensor chips with gas plasma-treated surface for the detection of hepatitis C virus core antigen. Coatings 2020; 10(8).
[http://dx.doi.org/10.3390/coatings10080753]

[124]   Zhang S, Garcia-D'Angeli A, Brennan JP, Huo Q. Predicting detection limits of enzyme-linked immunosorbent assay (ELISA) and bioanalytical techniques in general. Analyst (Lond) 2014; 139(2): 439-45.
[http://dx.doi.org/10.1039/C3AN01835K] [PMID: 24308031]

[125]   Lee S, Kim YS, Jo M, Jin M, Lee DK, Kim S. Chip-based detection of hepatitis C virus using RNA aptamers that specifically bind to HCV core antigen. Biochem Biophys Res Commun 2007; 358(1): 47-52.
[http://dx.doi.org/10.1016/j.bbrc.2007.04.057] [PMID: 17475212]

[126]   Uliana CV, Riccardi CS, Yamanaka H. Diagnostic tests for hepatitis C: recent trends in electrochemical immunosensor and genosensor analysis. World J Gastroenterol 2014; 20(42): 15476-91.
[http://dx.doi.org/10.3748/wjg.v20.i42.15476] [PMID: 25400433]

[127]   Shi S, Yu X, Gao Y, *et al.* Inhibition of hepatitis C virus production by aptamers against the core protein. J Virol 2014; 88(4): 1990-9.
[http://dx.doi.org/10.1128/JVI.03312-13] [PMID: 24307579]

[128]   Park JH, Jee MH, Kwon OS, Keum SJ, Jang SK. Infectivity of hepatitis C virus correlates with the amount of envelope protein E2: development of a new aptamer-based assay system suitable for measuring the infectious titer of HCV. Virology 2013; 439(1): 13-22.
[http://dx.doi.org/10.1016/j.virol.2013.01.014] [PMID: 23485371]

[129]   Chen F, Hu Y, Li D, Chen H, Zhang XL. CS-SELEX generates high-affinity ssDNA aptamers as molecular probes for hepatitis C virus envelope glycoprotein E2. PLoS One 2009; 4(12): e8142.
[http://dx.doi.org/10.1371/journal.pone.0008142] [PMID: 19997645]

[130]   Hwang KS, Lee SM, Eom K, *et al.* Nanomechanical microcantilever operated in vibration modes with use of RNA aptamer as receptor molecules for label-free detection of HCV helicase. Biosens Bioelectron 2007; 23(4): 459-65.
[http://dx.doi.org/10.1016/j.bios.2007.05.006] [PMID: 17616386]

[131]   Fatin MF, Rahim Ruslinda A, Gopinath SCB, Arshad MKM, Hashim U, Lakshmipriya T, *et al.* Co-ordinated split aptamer assembly and disassembly on Gold nanoparticle for functional detection of HIV-1 tat. Process Biochem 2019; 79: 32-9.
[http://dx.doi.org/10.1016/j.procbio.2018.12.016]

[132]   González VM, Martín ME, Fernández G, García-Sacristán A. Use of aptamers as diagnostics tools and antiviral agents for human viruses. Pharmaceuticals (Basel) 2016; 9(4): 1-34.
[http://dx.doi.org/10.3390/ph9040078] [PMID: 27999271]

[133]   Caglayan MO, Üstündağ Z. Spectrophotometric ellipsometry based Tat-protein RNA-aptasensor for HIV-1 diagnosis. Spectrochim Acta - Part A Mol Biomol Spectrosc 2020; 227(xxxx)

[134]   Susan L. Gillespie, Mary E.Paul, Javier Chinen WTS. . HIV infection and acquired immunodeficiency syndrome. In: Clinical Immunology 2013; 465-79.

[135] Tombelli S, Minunni M, Luzi E, Mascini M. Aptamer-based biosensors for the detection of HIV-1 Tat protein. Bioelectrochemistry 2005; 67(2 SPEC. ISS.): 135-41.

[136] Yamamoto R, Baba T, Kumar PK. Molecular beacon aptamer fluoresces in the presence of Tat protein of HIV-1. Genes Cells 2000; 5(5): 389-96.
[http://dx.doi.org/10.1046/j.1365-2443.2000.00331.x] [PMID: 10886366]

[137] Ugolini S, Mondor I, Parren PWHI, *et al.* Inhibition of virus attachment to CD4+ target cells is a major mechanism of T cell line-adapted HIV-1 neutralization. J Exp Med 1997; 186(8): 1287-98.
[http://dx.doi.org/10.1084/jem.186.8.1287] [PMID: 9334368]

[138] Davis KA, Lin Y, Abrams B, Jayasena SD. Staining of cell surface human CD4 with 2'-F-pyrimidi-e-containing RNA aptamers for flow cytometry. Nucleic Acids Res 1998; 26(17): 3915-24.
[http://dx.doi.org/10.1093/nar/26.17.3915] [PMID: 9705498]

[139] Kukushkin VI, Ivanov NM, Novoseltseva AA, *et al.* Highly sensitive detection of influenza virus with SERS aptasensor. PLoS One 2019; 14(4): e0216247.
[http://dx.doi.org/10.1371/journal.pone.0216247] [PMID: 31022287]

[140] Shiratori I, Akitomi J, Boltz DA, Horii K, Furuichi M, Waga I. Selection of DNA aptamers that bind to influenza A viruses with high affinity and broad subtype specificity. Biochem Biophys Res Commun 2014; 443(1): 37-41.
[http://dx.doi.org/10.1016/j.bbrc.2013.11.041] [PMID: 24269231]

[141] Bai C, Lu Z, Jiang H, *et al.* Aptamer selection and application in multivalent binding-based electrical impedance detection of inactivated H1N1 virus. Biosens Bioelectron 2018; 110(January): 162-7.
[http://dx.doi.org/10.1016/j.bios.2018.03.047] [PMID: 29609164]

[142] Kang J, Yeom G, Ha SJ, Kim MG. Development of a DNA aptamer selection method based on the heterogeneous sandwich form and its application in a colorimetric assay for influenza A virus detection. New J Chem 2019; 43(18): 6883-9.
[http://dx.doi.org/10.1039/C8NJ06458J]

[143] Kim SM, Kim J, Noh S, Sohn H, Lee T. Recent development of aptasensor for influenza virus detection. Biochip J 2020; 14(4): 1-13.
[http://dx.doi.org/10.1007/s13206-020-4401-2] [PMID: 33224441]

[144] Kwon J, Lee Y, Lee T, Ahn JH. Aptamer-based field-effect transistor for detection of avian influenza virus in chicken serum. Anal Chem 2020; 92(7): 5524-31.
[http://dx.doi.org/10.1021/acs.analchem.0c00348] [PMID: 32148026]

[145] Chen H, Park SG, Choi N, *et al.* SERS imaging-based aptasensor for ultrasensitive and reproducible detection of influenza virus A. Biosens Bioelectron 2020; 167(August): 112496.
[http://dx.doi.org/10.1016/j.bios.2020.112496] [PMID: 32818752]

[146] Kang J, Yeom G, Jang H, Park CJ, Kim MG. Highly sensitive and universal detection strategy based on a colorimetric assay using target-specific heterogeneous sandwich DNA aptamer. Anal Chim Acta 2020; 1123: 73-80.
[http://dx.doi.org/10.1016/j.aca.2020.05.012] [PMID: 32507242]

[147] Bhardwaj J, Chaudhary N, Kim H, Jang J. Subtyping of influenza A H1N1 virus using a label-free electrochemical biosensor based on the DNA aptamer targeting the stem region of HA protein. Anal Chim Acta 2019; 1064: 94-103.
[http://dx.doi.org/10.1016/j.aca.2019.03.005] [PMID: 30982523]

[148] Suenaga E, Kumar PKR. An aptamer that binds efficiently to the hemagglutinins of highly pathogenic avian influenza viruses (H5N1 and H7N7) and inhibits hemagglutinin-glycan interactions. Acta Biomater 2014; 10(3): 1314-23.4.

[149] Gopinath SCB, Misono TS, Kawasaki K, *et al.* An RNA aptamer that distinguishes between closely related human influenza viruses and inhibits haemagglutinin-mediated membrane fusion. J Gen Virol 2006; 87(Pt 3): 479-87.

[http://dx.doi.org/10.1099/vir.0.81508-0] [PMID: 16476969]

[150] Gopinath SCB, Kumar PKR. Aptamers that bind to the hemagglutinin of the recent pandemic influenza virus H1N1 and efficiently inhibit agglutination. Acta Biomater 2013; 9(11): 8932-41.
[http://dx.doi.org/10.1016/j.actbio.2013.06.016] [PMID: 23791676]

[151] Wang R, Zhao J, Jiang T, *et al.* Selection and characterization of DNA aptamers for use in detection of avian influenza virus H5N1. J Virol Methods 2013; 189(2): 362-9.
[http://dx.doi.org/10.1016/j.jviromet.2013.03.006] [PMID: 23523887]

[152] Nguyen VT, Seo HB, Kim BC, Kim SK, Song CS, Gu MB. Highly sensitive sandwich-type SPR based detection of whole H5Nx viruses using a pair of aptamers. Biosens Bioelectron 2016; 86: 293-300.
[http://dx.doi.org/10.1016/j.bios.2016.06.064] [PMID: 27387259]

[153] Lai HC, Wang CH, Liou TM, Lee GB. Influenza A virus-specific aptamers screened by using an integrated microfluidic system. Lab Chip 2014; 14(12): 2002-13.
[http://dx.doi.org/10.1039/C4LC00187G] [PMID: 24820138]

[154] Kim SH, Lee J, Lee BH, Song CS, Gu MB. Specific detection of avian influenza H5N2 whole virus particles on lateral flow strips using a pair of sandwich-type aptamers. Biosens Bioelectron 2019; 134(April): 123-9.
[http://dx.doi.org/10.1016/j.bios.2019.03.061] [PMID: 30986614]

[155] Bruno JG, Carrillo MP, Richarte AM, Phillips T, Andrews C, Lee JS. Development, screening, and analysis of DNA aptamer libraries potentially useful for diagnosis and passive immunity of arboviruses. BMC Res Notes 2012; 5: 633.
[http://dx.doi.org/10.1186/1756-0500-5-633] [PMID: 23148669]

[156] Shi P-Y. Molecular Virology and Control of Flaviviruses. Caister Academic Press 2012.

[157] Basso CR, Crulhas BP, Magro M, Vianello F, Pedrosa VA. A new immunoassay of hybrid nanomater conjugated to aptamers for the detection of dengue virus. Talanta 2019; 197: 482-90.
[http://dx.doi.org/10.1016/j.talanta.2019.01.058] [PMID: 30771965]

[158] Messina JP, Brady OJ, Scott TW, *et al.* Global spread of dengue virus types: mapping the 70 year history. Trends Microbiol 2014; 22(3): 138-46.
[http://dx.doi.org/10.1016/j.tim.2013.12.011] [PMID: 24468533]

[159] Lee KH, Zeng H. Aptamer-based elisa assay for highly specific and sensitive detection of zika ns1 protein. Anal Chem 2017; 89(23): 12743-8.
[http://dx.doi.org/10.1021/acs.analchem.7b02862] [PMID: 29120623]

[160] Alves LN, Abalo AA, Argondizzo APC, Rocha HF, Silva D, Cortez CM, *et al.* Selection and evaluation of the binding of aptamers against NS5 Zika virus using fluorescence spectroscopy. AIP Conf Proc 2018; 2040(November): 1-5.
[http://dx.doi.org/10.1063/1.5079160]

[161] Ellenbecker M, Sears L, Li P, Lanchy JM, Lodmell JS. Characterization of RNA aptamers directed against the nucleocapsid protein of Rift Valley fever virus. Antiviral Res 2012; 93(3): 330-9.
[http://dx.doi.org/10.1016/j.antiviral.2012.01.002] [PMID: 22252167]

[162] Kondratov IG, Khasnatinov MA, Potapova UV, *et al.* Obtaining aptamers to a fragment of surface protein E of tick-borne encephalitis virus. Dokl Biochem Biophys 2013; 448(1): 19-21.
[http://dx.doi.org/10.1134/S1607672913010067] [PMID: 23478980]

[163] Saraf N, Villegas M, Willenberg BJ, Seal S. Multiplex viral detection platform based on a aptamers-integrated microfluidic channel. ACS Omega 2019; 4(1): 2234-40.
[http://dx.doi.org/10.1021/acsomega.8b03277] [PMID: 30729227]

[164] Jang KJ, Lee NR, Yeo WS, Jeong YJ, Kim DE. Isolation of inhibitory RNA aptamers against severe acute respiratory syndrome (SARS) coronavirus NTPase/Helicase. Biochem Biophys Res Commun 2008; 366(3): 738-44.
[http://dx.doi.org/10.1016/j.bbrc.2007.12.020] [PMID: 18082623]

[165]   Ahn DG, Jeon IJ, Kim JD, *et al.* RNA aptamer-based sensitive detection of SARS coronavirus nucleocapsid protein. Analyst (Lond) 2009; 134(9): 1896-901.
[http://dx.doi.org/10.1039/b906788d] [PMID: 19684916]

[166]   Cho SJ, Woo HM, Kim KS, Oh JW, Jeong YJ. Novel system for detecting SARS coronavirus nucleocapsid protein using an ssDNA aptamer. J Biosci Bioeng 2011; 112(6): 535-40.
[http://dx.doi.org/10.1016/j.jbiosc.2011.08.014] [PMID: 21920814]

[167]   Hong SL, Xiang MQ, Tang M, Pang DW, Zhang ZL. Ebola virus aptamers: from highly efficient selection to application on magnetism-controlled chips. Anal Chem 2019; 91(5): 3367-73.
[http://dx.doi.org/10.1021/acs.analchem.8b04623] [PMID: 30740973]

[168]   Zhang L, Fang X, Liu X, *et al.* Discovery of sandwich type COVID-19 nucleocapsid protein DNA aptamers. Chem Commun (Camb) 2020; 56(70): 10235-8.
[http://dx.doi.org/10.1039/D0CC03993D] [PMID: 32756614]

[169]   Uhteg K, Jarrett J, Richards M, Howard C, Morehead E, Geahr M, *et al.* Since January 2020 Elsevier has created a COVID-19 resource centre with free information in English and Mandarin on the novel coronavirus COVID- 19 2020. January

[170]   D'Cruz RJ, Currier AW, Sampson VB. Laboratory testing methods for novel severe acute respiratory syndrome-coronavirus-2 (SARS-CoV-2). Front Cell Dev Biol 2020; 8(June): 468.
[http://dx.doi.org/10.3389/fcell.2020.00468] [PMID: 32582718]

[171]   Yan C, Cui J, Huang L, Du B, Chen L, Xue G, *et al.* Since January 2020 Elsevier has created a COVID-19 resource centre with free information in English and Mandarin on the novel coronavirus COVID- 19 2020. January

[172]   Baek YH, Um J, Antigua KJC, *et al.* Development of a reverse transcription-loop-mediated isothermal amplification as a rapid early-detection method for novel SARS-CoV-2. Emerg Microbes Infect 2020; 9(1): 998-1007.
[http://dx.doi.org/10.1080/22221751.2020.1756698] [PMID: 32306853]

[173]   Lu R, Wu X, Wan Z, *et al.* Development of a novel reverse transcription loop-mediated isothermal amplification method for rapid detection of SARS-CoV-2. Virol Sin 2020; 35(3): 344-7.
[http://dx.doi.org/10.1007/s12250-020-00218-1] [PMID: 32239445]

[174]   Zhang Y, Odiwuor N, Xiong J, Sun L, Nyaruaba RO, Wei H, *et al.* Rapid molecular detection of SARS-CoV-2 (COVID-19) virus RNA using colorimetric LAMP. medRxiv 2020; 2

[175]   Broughton JP, Deng X, Yu G, *et al.* CRISPR-Cas12-based detection of SARS-CoV-2. Nat Biotechnol 2020; 38(7): 870-4.
[http://dx.doi.org/10.1038/s41587-020-0513-4] [PMID: 32300245]

[176]   Okba NMA, Müller MA, Li W, Wang C. SARS-CoV-2 specific antibody responses in COVID-19 patients. medRxiv 2020; 1-18.

[177]   Hoffman T, Nissen K, Krambrich J, *et al.* Evaluation of a COVID-19 IgM and IgG rapid test; an efficient tool for assessment of past exposure to SARS-CoV-2. Infect Ecol Epidemiol 2020; 10(1): 1754538.
[http://dx.doi.org/10.1080/20008686.2020.1754538] [PMID: 32363011]

[178]   Che XY, Hao W, Wang Y, *et al.* Nucleocapsid protein as early diagnostic marker for SARS. Emerg Infect Dis 2004; 10(11): 1947-9.
[http://dx.doi.org/10.3201/eid1011.040516] [PMID: 15550204]

[179]   Li YH, Li J, Liu XE, *et al.* Detection of the nucleocapsid protein of severe acute respiratory syndrome coronavirus in serum: comparison with results of other viral markers. J Virol Methods 2005; 130(1-2): 45-50.
[http://dx.doi.org/10.1016/j.jviromet.2005.06.001] [PMID: 16024098]

[180]   Lau SKP, Woo PCY, Wong BHL, *et al.* Detection of severe acute respiratory syndrome (SARS)

coronavirus nucleocapsid protein in sars patients by enzyme-linked immunosorbent assay. J Clin Microbiol 2004; 42(7): 2884-9.
[http://dx.doi.org/10.1128/JCM.42.7.2884-2889.2004] [PMID: 15243033]

[181] Chen Z, Wu Q, Chen J, Ni X, Dai J. A DNA aptamer based method for detection of SARS-COV-2 nucleocapsid protein. Virol Sin 2020; 35(3): 351-4.
[http://dx.doi.org/10.1007/s12250-020-00236-z] [PMID: 32451881]

[182] Song Y, Song J, Wei X, *et al.* Discovery of aptamers targeting the receptor-binding domain of the sars-cov-2 spike glycoprotein. Anal Chem 2020; 92(14): 9895-900.
[http://dx.doi.org/10.1021/acs.analchem.0c01394] [PMID: 32551560]

[183] Chen H, Zhang Q, Qiao L, *et al.* Cdc6 contributes to abrogating the G1 checkpoint under hypoxic conditions in HPV E7 expressing cells. Sci Rep 2017; 7(1): 2927.
[http://dx.doi.org/10.1038/s41598-017-03060-w] [PMID: 28592805]

[184] Toscano-Garibay JD, Benítez-Hess ML, Alvarez-Salas LM. Isolation and characterization of an RNA aptamer for the HPV-16 E7 oncoprotein. Arch Med Res 2011; 42(2): 88-96.
[http://dx.doi.org/10.1016/j.arcmed.2011.02.005] [PMID: 21565620]

[185] Aspermair P, Mishyn V, Bintinger J, *et al.* Reduced graphene oxide-based field effect transistors for the detection of E7 protein of human papillomavirus in saliva. Anal Bioanal Chem 2021; 413(3): 779-87.
[http://dx.doi.org/10.1007/s00216-020-02879-z] [PMID: 32816088]

[186] Butz K, Denk C, Ullmann A, Scheffner M, Hoppe-Seyler F. Induction of apoptosis in human papillomaviruspositive cancer cells by peptide aptamers targeting the viral E6 oncoprotein. Proc Natl Acad Sci USA 2000; 97(12): 6693-7.
[http://dx.doi.org/10.1073/pnas.110538897] [PMID: 10829072]

[187] Gopinath SCB, Hayashi K, Kumar PKR. Aptamer that binds to the gD protein of herpes simplex virus 1 and efficiently inhibits viral entry. J Virol 2012; 86(12): 6732-44.
[http://dx.doi.org/10.1128/JVI.00377-12] [PMID: 22514343]

[188] Kumar N, Sood D, Singh S, Kumar S, Chandra R. High bio-recognizing aptamer designing and optimization against human herpes virus-5. Eur J Pharm Sci 2021; 156: 105572.
[http://dx.doi.org/10.1016/j.ejps.2020.105572]

[189] Férir G, Gordts SC, Schols D. HIV-1 and its resistance to peptidic carbohydrate-binding agents (CBAs): an overview. Molecules 2014; 19(12): 21085-112.
[http://dx.doi.org/10.3390/molecules191221085] [PMID: 25517345]

[190] Acquah C, Jeevanandam J, Tan KX, Danquah MK. Engineered aptamers for enhanced COVID-19 theranostics. Cell Mol Bioeng 2021; 1-13.
[PMID: 33488836]

[191] Hu WS, Hughes SH. HIV-1 reverse transcription. Cold Spring Harb Perspect Med 2012; 2(10): a006882.
[http://dx.doi.org/10.1101/cshperspect.a006882] [PMID: 23028129]

[192] DeStefano JJ, Nair GR. Novel aptamer inhibitors of human immunodeficiency virus reverse transcriptase. Oligonucleotides 2008; 18(2): 133-44.
[http://dx.doi.org/10.1089/oli.2008.0103] [PMID: 18637731]

[193] Nguyen PDM, Zheng J, Gremminger TJ, *et al.* Binding interface and impact on protease cleavage for an RNA aptamer to HIV-1 reverse transcriptase. Nucleic Acids Res 2020; 48(5): 2709-22.
[http://dx.doi.org/10.1093/nar/gkz1224] [PMID: 31943114]

[194] Whatley AS, Ditzler MA, Lange MJ, *et al.* Potent inhibition of HIV-1 reverse transcriptase and replication by nonpseudoknot,"UCAA-motif" RNA aptamers. Mol Ther Nucleic Acids 2013; 2: e71.
[http://dx.doi.org/10.1038/mtna.2012.62] [PMID: 23385524]

[195] Lange MJ, Nguyen PDM, Callaway MK, Johnson MC, Burke DH. RNA-protein interactions govern

antiviral specificity and encapsidation of broad spectrum anti-HIV reverse transcriptase aptamers. Nucleic Acids Res 2017; 45(10): 6087-97.
[http://dx.doi.org/10.1093/nar/gkx155] [PMID: 28334941]

[196] Tuerk C, MacDougal S, Gold L. RNA pseudoknots that inhibit human immunodeficiency virus type 1 reverse transcriptase. Proc Natl Acad Sci USA 1992; 89(15): 6988-92.
[http://dx.doi.org/10.1073/pnas.89.15.6988] [PMID: 1379730]

[197] Kensch O, Connolly BA, Steinhoff HJ, McGregor A, Goody RS, Restle T. HIV-1 reverse transcriptase-pseudoknot RNA aptamer interaction has a binding affinity in the low picomolar range coupled with high specificity. J Biol Chem 2000; 275(24): 18271-8.
[http://dx.doi.org/10.1074/jbc.M001309200] [PMID: 10751399]

[198] Schneider DJ, Feigon J, Hostomsky Z, Gold L. High-affinity ssDNA inhibitors of the reverse transcriptase of type 1 human immunodeficiency virus. Biochemistry 1995; 34(29): 9599-610.
[http://dx.doi.org/10.1021/bi00029a037] [PMID: 7542922]

[199] Kissel JD, Held DM, Hardy RW, Burke DH. Single-stranded DNA aptamer RT1t49 inhibits RT polymerase and RNase H functions of HIV type 1, HIV type 2, and SIVCPZ RTs. AIDS Res Hum Retroviruses 2007; 23(5): 699-708.
[http://dx.doi.org/10.1089/aid.2006.0262] [PMID: 17530996]

[200] Miller MT, Tuske S, Das K, DeStefano JJ, Arnold E. Structure of HIV-1 reverse transcriptase bound to a novel 38-mer hairpin template-primer DNA aptamer. Protein Sci 2016; 25(1): 46-55.
[http://dx.doi.org/10.1002/pro.2776] [PMID: 26296781]

[201] Somasunderam A, Ferguson MR, Rojo DR, *et al.* Combinatorial selection, inhibition, and antiviral activity of DNA thioaptamers targeting the RNase H domain of HIV-1 reverse transcriptase. Biochemistry 2005; 44(30): 10388-95.
[http://dx.doi.org/10.1021/bi0507074] [PMID: 16042416]

[202] Cherepanov P, Esté JA, Rando RF, *et al.* Mode of interaction of G-quartets with the integrase of human immunodeficiency virus type 1. Mol Pharmacol 1997; 52(5): 771-80.
[http://dx.doi.org/10.1124/mol.52.5.771] [PMID: 9351967]

[203] Rivieccio E, Tartaglione L, Esposito V, *et al.* Structural studies and biological evaluation of T30695 variants modified with single chiral glycerol-T reveal the importance of LEDGF/p75 for the aptamer anti-HIV-integrase activities. Biochim Biophys Acta, Gen Subj 2019; 1863(2): 351-61.
[http://dx.doi.org/10.1016/j.bbagen.2018.11.001] [PMID: 30414444]

[204] de Soultrait VR, Lozach PY, Altmeyer R, Tarrago-Litvak L, Litvak S, Andréola ML. DNA aptamers derived from HIV-1 RNase H inhibitors are strong anti-integrase agents. J Mol Biol 2002; 324(2): 195-203.
[http://dx.doi.org/10.1016/S0022-2836(02)01064-1] [PMID: 12441099]

[205] Virgilio A, Amato T, Petraccone L, *et al.* Improvement of the activity of the anti-HIV-1 integrase aptamer T30175 by introducing a modified thymidine into the loops. Sci Rep 2018; 8(1): 7447.
[http://dx.doi.org/10.1038/s41598-018-25720-1] [PMID: 29749406]

[206] Faure-Perraud A, Métifiot M, Reigadas S, *et al.* The guanine-quadruplex aptamer 93del inhibits HIV-1 replication ex vivo by interfering with viral entry, reverse transcription and integration. Antivir Ther 2011; 16(3): 383-94.
[http://dx.doi.org/10.3851/IMP1756] [PMID: 21555821]

[207] Phan AT, Kuryavyi V, Ma JB, Faure A, Andréola ML, Patel DJ. An interlocked dimeric parallel-stranded DNA quadruplex: a potent inhibitor of HIV-1 integrase. Proc Natl Acad Sci USA 2005; 102(3): 634-9.
[http://dx.doi.org/10.1073/pnas.0406278102] [PMID: 15637158]

[208] Liu Y, Zhang Y, Ye G, Yang Z, Zhang L, Zhang L. *In vitro* selection of G-rich RNA aptamers that target HIV-1 integrase. Sci China B Chem 2008; 51(5): 401-13.
[http://dx.doi.org/10.1007/s11426-008-0056-x]

[209]   Rose KM, Alves Ferreira-Bravo I, Li M, *et al.* Selection of 2′-deoxy-2′-fluoroarabino nucleic acid (FANA) aptamers that bind HIV-1 integrase with picomolar affinity. ACS Chem Biol 2019; 14(10): 2166-75.
        [http://dx.doi.org/10.1021/acschembio.9b00237] [PMID: 31560515]

[210]   Symensma TL, Giver L, Zapp M, Takle GB, Ellington AD. RNA aptamers selected to bind human immunodeficiency virus type 1 Rev *in vitro* are Rev responsive *in vivo.* J Virol 1996; 70(1): 179-87.
        [http://dx.doi.org/10.1128/jvi.70.1.179-187.1996] [PMID: 8523524]

[211]   Dearborn AD, Eren E, Watts NR, *et al.* Structure of an RNA aptamer that can inhibit HIV-1 by blocking Rev-cognate RNA (RRE) binding and Rev-Rev association. Structure 2018; 26(9): 1187-1195.e4.
        [http://dx.doi.org/10.1016/j.str.2018.06.001] [PMID: 30017564]

[212]   Kwong PD, Wyatt R, Robinson J, Sweet RW, Sodroski J, Hendrickson WA. Structure of an HIV gp120 envelope glycoprotein in complex with the CD4 receptor and a neutralizing human antibody. Nature 1998; 393(6686): 648-59.
        [http://dx.doi.org/10.1038/31405] [PMID: 9641677]

[213]   Sayer N, Ibrahim J, Turner K, Tahiri-Alaoui A, James W. Structural characterization of a 2'F-RNA aptamer that binds a HIV-1 SU glycoprotein, gp120. Biochem Biophys Res Commun 2002; 293(3): 924-31.
        [http://dx.doi.org/10.1016/S0006-291X(02)00308-X] [PMID: 12051747]

[214]   Khati M, Schüman M, Ibrahim J, Sattentau Q, Gordon S, James W. Neutralization of infectivity of diverse R5 clinical isolates of human immunodeficiency virus type 1 by gp120-binding 2'F-RNA aptamers. J Virol 2003; 77(23): 12692-8.
        [http://dx.doi.org/10.1128/JVI.77.23.12692-12698.2003] [PMID: 14610191]

[215]   Mufhandu HT, Gray ES, Madiga MC, *et al.* UCLA1, a synthetic derivative of a gp120 RNA aptamer, inhibits entry of human immunodeficiency virus type 1 subtype C. J Virol 2012; 86(9): 4989-99.
        [http://dx.doi.org/10.1128/JVI.06893-11] [PMID: 22379083]

[216]   Sepehri Zarandi H, Behbahani M, Mohabatkar H. *In Silico* Selection of Gp120 ssDNA Aptamer to HIV-1 SLAS Discov 2020; 25(9): 1087-93.

[217]   Virgilio A, Esposito V, Tassinari M, Nadai M, Richter SN, Galeone A. Novel monomolecular derivatives of the anti-HIV-1 G-quadruplex-forming Hotoda's aptamer containing inversion of polarity sites. Eur J Med Chem 2020; 208: 112786.
        [http://dx.doi.org/10.1016/j.ejmech.2020.112786] [PMID: 32911256]

[218]   Lochrie MA, Waugh S, Pratt DG Jr, Clever J, Parslow TG, Polisky B. *In vitro* selection of RNAs that bind to the human immunodeficiency virus type-1 gag polyprotein. Nucleic Acids Res 1997; 25(14): 2902-10.
        [http://dx.doi.org/10.1093/nar/25.14.2902] [PMID: 9207041]

[219]   Wills JW, Craven RC. Form, function, and use of retroviral gag proteins. AIDS 1991; 5(6): 639-54.
        [http://dx.doi.org/10.1097/00002030-199106000-00002] [PMID: 1883539]

[220]   Ramalingam D, Duclair S, Datta SA, Ellington A, Rein A, Prasad VR. RNA aptamers directed to human immunodeficiency virus type 1 Gag polyprotein bind to the matrix and nucleocapsid domains and inhibit virus production. J Virol 2011; 85(1): 305-14.
        [http://dx.doi.org/10.1128/JVI.02626-09] [PMID: 20980522]

[221]   Gu M, Rice CM. Structures of hepatitis C virus nonstructural proteins required for replicase assembly and function. Curr Opin Virol 2013; 3(2): 129-36.
        [http://dx.doi.org/10.1016/j.coviro.2013.03.013] [PMID: 23601958]

[222]   Bellecave P, Andreola ML, Ventura M, Tarrago-Litvak L, Litvak S, Astier-Gin T. Selection of DNA aptamers that bind the RNA-dependent RNA polymerase of hepatitis C virus and inhibit viral RNA synthesis *in vitro.* Oligonucleotides 2003; 13(6): 455-63.

[http://dx.doi.org/10.1089/154545703322860771] [PMID: 15025912]

[223] Kanamori H, Yuhashi K, Uchiyama Y, Kodama T, Ohnishi S. *In vitro* selection of RNA aptamers that bind the RNA-dependent RNA polymerase of hepatitis C virus: a possible role of GC-rich RNA motifs in NS5B binding. Virology 2009; 388(1): 91-102.
[http://dx.doi.org/10.1016/j.virol.2009.02.032] [PMID: 19328515]

[224] Friebe P, Boudet J, Simorre JP, Bartenschlager R. Kissing-loop interaction in the 3' end of the hepatitis C virus genome essential for RNA replication. J Virol 2005; 79(1): 380-92.
[http://dx.doi.org/10.1128/JVI.79.1.380-392.2005] [PMID: 15596831]

[225] Zhang J, Yamada O, Sakamoto T, *et al.* Inhibition of hepatitis C virus replication by pol III-directed overexpression of RNA decoys corresponding to stem-loop structures in the NS5B coding region. Virology 2005; 342(2): 276-85.
[http://dx.doi.org/10.1016/j.virol.2005.08.003] [PMID: 16139319]

[226] Lee CH, Lee YJ, Kim JH, *et al.* Inhibition of hepatitis C virus (HCV) replication by specific RNA aptamers against HCV NS5B RNA replicase. J Virol 2013; 87(12): 7064-74.
[http://dx.doi.org/10.1128/JVI.00405-13] [PMID: 23596299]

[227] Lee CH, Lee SH, Kim JH, Noh YH, Noh GJ, Lee SW. Pharmacokinetics of a cholesterol-conjugated aptamer against the hepatitis C virus (HCV) NS5B protein. Mol Ther Nucleic Acids 2015; 4: e254.
[http://dx.doi.org/10.1038/mtna.2015.30] [PMID: 26440598]

[228] Yu X, Gao Y, Xue B, *et al.* Inhibition of hepatitis C virus infection by NS5A-specific aptamer. Antiviral Res 2014; 106: 116-24.
[http://dx.doi.org/10.1016/j.antiviral.2014.03.020] [PMID: 24713119]

[229] Urvil PT, Kakiuchi N, Zhou DM, Shimotohno K, Kumar PK, Nishikawa S. Selection of RNA aptamers that bind specifically to the NS3 protease of hepatitis C virus. Eur J Biochem 1997; 248(1): 130-8.
[http://dx.doi.org/10.1111/j.1432-1033.1997.t01-1-00130.x] [PMID: 9310370]

[230] Fukuda K, Vishnuvardhan D, Sekiya S, *et al.* Isolation and characterization of RNA aptamers specific for the hepatitis C virus nonstructural protein 3 protease. Eur J Biochem 2000; 267(12): 3685-94.
[http://dx.doi.org/10.1046/j.1432-1327.2000.01400.x] [PMID: 10848986]

[231] Nishikawa F, Kakiuchi N, Funaji K, Fukuda K, Sekiya S, Nishikawa S. Inhibition of HCV NS3 protease by RNA aptamers in cells. Nucleic Acids Res 2003; 31(7): 1935-43.
[http://dx.doi.org/10.1093/nar/gkg291] [PMID: 12655010]

[232] Hwang B, Cho JS, Yeo HJ, *et al.* Isolation of specific and high-affinity RNA aptamers against NS3 helicase domain of hepatitis C virus. RNA 2004; 10(8): 1277-90.
[http://dx.doi.org/10.1261/rna.7100904] [PMID: 15247433]

[233] Gao Y, Yu X, Xue B, *et al.* Inhibition of hepatitis C virus infection by DNA aptamer against NS2 protein. PLoS One 2014; 9(2): e90333.
[http://dx.doi.org/10.1371/journal.pone.0090333] [PMID: 24587329]

[234] Chen F, Chen SC, Zhou J, Chen ZD, Chen F. Identification of aptamer-binding sites in hepatitis C virus envelope glycoprotein e2. Iran J Med Sci 2015; 40(1): 63-7.
[PMID: 25648186]

[235] Yang D, Meng X, Yu Q, *et al.* Inhibition of hepatitis C virus infection by DNA aptamer against envelope protein. Antimicrob Agents Chemother 2013; 57(10): 4937-44.
[http://dx.doi.org/10.1128/AAC.00897-13] [PMID: 23877701]

[236] Zhang Z, Zhang J, Pei X, *et al.* An aptamer targets HBV core protein and suppresses HBV replication in HepG2.2.15 cells. Int J Mol Med 2014; 34(5): 1423-9.
[http://dx.doi.org/10.3892/ijmm.2014.1908] [PMID: 25174447]

[237] Butz K, Denk C, Fitscher B, *et al.* Peptide aptamers targeting the hepatitis B virus core protein: a new class of molecules with antiviral activity. Oncogene 2001; 20(45): 6579-86.

[http://dx.doi.org/10.1038/sj.onc.1204805] [PMID: 11641783]

[238]   Orabi A, Bieringer M, Geerlof A, Bruss V. An aptamer against the matrix binding domain on the hepatitis B virus capsid impairs virion formation. J Virol 2015; 89(18): 9281-7.
[http://dx.doi.org/10.1128/JVI.00466-15] [PMID: 26136564]

[239]   Liu J, Yang Y, Hu B, *et al.* Development of HBsAg-binding aptamers that bind HepG2.2.15 cells *via* HBV surface antigen. Virol Sin 2010; 25(1): 27-35.
[http://dx.doi.org/10.1007/s12250-010-3091-7] [PMID: 20960281]

[240]   Shum KT, Tanner JA. Differential inhibitory activities and stabilisation of DNA aptamers against the SARS coronavirus helicase. ChemBioChem 2008; 9(18): 3037-45.
[http://dx.doi.org/10.1002/cbic.200800491] [PMID: 19031435]

[241]   Parashar NC, Poddar J, Chakrabarti S, Parashar G. Repurposing of SARS-CoV nucleocapsid protein specific nuclease resistant RNA aptamer for therapeutics against SARS-CoV-2. Infect Genet Evol 2020; 85: 104497.
[http://dx.doi.org/10.1016/j.meegid.2020.104497] [PMID: 32791240]

[242]   Wallukat G, Müller J, Haberland A, *et al.* Aptamer BC007 for neutralization of pathogenic autoantibodies directed against G-protein coupled receptors: A vision of future treatment of patients with cardiomyopathies and positivity for those autoantibodies. Atherosclerosis 2016; 244: 44-7.
[http://dx.doi.org/10.1016/j.atherosclerosis.2015.11.001] [PMID: 26584137]

[243]   Weisshoff H, Krylova O, Nikolenko H, *et al.* Aptamer BC 007 - Efficient binder of spreading-crucial SARS-CoV-2 proteins. Heliyon 2020; 6(11): e05421.
[http://dx.doi.org/10.1016/j.heliyon.2020.e05421] [PMID: 33163683]

[244]   Schmitz A, Weber A, Bayin M, Breuers S, Famulok M, Mayer GA. SARS-CoV-2 spike binding DNA aptamer that inhibits pseudovirus infection *in vitro* by an RBD independent mechanism. bioRxiv 2020.

[245]   Sriwilaijaroen N, Wilairat P, Hiramatsu H, *et al.* Mechanisms of the action of povidone-iodine against human and avian influenza A viruses: its effects on hemagglutination and sialidase activities. Virol J 2009; 6(1): 124.
[http://dx.doi.org/10.1186/1743-422X-6-124] [PMID: 19678928]

[246]   Shum KT, Zhou J, Rossi JJ. Aptamer-based therapeutics: new approaches to combat human viral diseases. Pharmaceuticals (Basel) 2013; 6(12): 1507-42.
[http://dx.doi.org/10.3390/ph6121507] [PMID: 24287493]

[247]   Li W, Feng X, Yan X, Liu K, Deng L. A DNA aptamer against influenza a virus: an effective inhibitor to the hemagglutinin-glycan interactions. Nucleic Acid Ther 2016; 26(3): 166-72.
[http://dx.doi.org/10.1089/nat.2015.0564] [PMID: 26904922]

[248]   Cheng C, Dong J, Yao L, *et al.* Potent inhibition of human influenza H5N1 virus by oligonucleotides derived by SELEX. Biochem Biophys Res Commun 2008; 366(3): 670-4.
[http://dx.doi.org/10.1016/j.bbrc.2007.11.183] [PMID: 18078808]

[249]   Kwon HM, Lee KH, Han BW, Han MR, Kim DH, Kim DE. An RNA aptamer that specifically binds to the glycosylated hemagglutinin of avian influenza virus and suppresses viral infection in cells. PLoS One 2014; 9(5): e97574.
[http://dx.doi.org/10.1371/journal.pone.0097574] [PMID: 24835440]

[250]   Yuan S, Zhang N, Singh K, *et al.* Cross-protection of influenza A virus infection by a DNA aptamer targeting the PA endonuclease domain. Antimicrob Agents Chemother 2015; 59(7): 4082-93.
[http://dx.doi.org/10.1128/AAC.00306-15] [PMID: 25918143]

CHAPTER 2

Host-Directed, Antibiotic-Adjuvant Combination, and Antibiotic-Antibiotic Combination for Treating Multidrug-Resistant (MDR) Gram-Negative Pathogens

Wattana Leowattana[1,*], **Pathomthep Leowattana**[2] and **Tawithep Leowattana**[3]

[1] *Department of Clinical Tropical Medicine, Faculty of Tropical Medicine, Mahidol University, 420/6 Rajavithi road, Rachatawee, Bangkok 10400, Thailand*

[2] *Tivanon Medical Clinics,99 Tivanon Road, Muang, Nonthaburi 11000, Thailand*

[3] *Department of Medicine, Faculty of Medicine, Srinakharinwirot University,114 Sukhumvit 23, Wattana District, Bangkok 10110, Thailand*

Abstract: Antibiotics were firstly used for the treatment of critical infections in the 1940s. They could save patients' lives and increased life spans by improving the outcome of serious infections. Antibiotics are the most commonly used drugs in a healthcare environment. However, antibiotics are not correctly prescribed, due to improper antibiotic selection, not suitable dosing, inappropriate treatment duration, and wrong treatment in nonbacterial conditions. Consequently, the rapid emergence of resistant bacteria occurs worldwide because they could adapt and compete with environmental stress. There are 4 primary mechanisms of resistance to counter the antibiotics: (i) modification of the target, (ii) enzymatic inhibition of the antibiotics, (iii) active efflux of the antibiotics, and (iv) changing membrane permeability. Carbapenem-resistant *Pseudomonas aeruginosa* (CR-PA), CR *Acinetobacter baumannii* (CR-AB), and CR *Enterobacteriaceae* were declared by World Health Organization (WHO) as the three most important pathogens that pose the greatest threat to human health. Moreover, they are also multidrug-resistant (MDR) and usually resist almost all of the most effective antibiotics, including carbapenem and fourth-generation cephalosporin. There is an urgent need to develop new strategies to combat antibiotic resistance and preserve the existing antibiotics. Numerous approaches, including host-directed, antibiotic-adjuvant combination, and antibiotic-antibiotic combination therapy, have been put forward to bring about antibiotic efficacy against MDR pathogens.

Keywords: Antibiotic-Adjuvant Combination, Antibiotic-Antibiotic Combination, Gram-Negative Pathogens, Host-Directed, Multidrug-Resistant.

---

* **Corresponding author Wattana Leowattana:** Department of Clinical Tropical Medicine, Faculty of Tropical Medicine, Mahidol University, Bangkok, Thailand; Tel: +6623549100, Ext. 3168; Fax: +6623549168; E-mail:wattana.leo@mahidol.ac.th

Atta-ur-Rahman, *FRS* (Ed.)

INTRODUCTION

Bacterial infections remain a significant global health problem. Antibiotics have bring about a disease in prevalence from bacterial pathogens; however, we increasingly confront antimicrobial resistance (AMR) to these antibiotics, and the development of new antibiotics lags beyond the emergence of the bacterial resistance. The problem is not confined to human-linked habitats since different works have shown that other ecosystems, including animals, soil, and water bodies, contribute to the origin, spread, and maintenance of AMR. Moreover, multidrug-resistant (MDR) microorganisms are among the most prominent healthcare problems of the 21st century and are responsible for 60,000 - 70,000 deaths per year in the United States and Europe [1, 2]. These situations are prevalent in the low- and middle-income countries where the resistant strains are hard to detect. Recently, World Health Organization (WHO) has reported that approximately 51% and 65% of infections are resistant to penicillin and ciprofloxacin, respectively, in many countries [3]. The antibiotics are widely used in medical therapeutic facilities and selection pressure is developed that increases MDR pathogens; thus there is a further need for new-generation antibiotics with novel antibacterial properties [4, 5]. To mitigate the MDR pathogen, combination treatment is an urgent option. Combining two antibiotics comprised of a drug targeting the antibiotic resistance activity and an adjuvant are promising new therapeutic strategies [6, 7]. Furthermore, the clinical manifestation of bacterial pathogens reflects a complex interplay between the host, pathogen, and antibiotics. Owing to the innate immune response playing a crucial role in combating bacterial infection, a host-directed therapy combined with an appropriate antimicrobial agent may reduce the antibacterial resistance. This approach may led to a successful clinical outcome and resolve antimicrobial-resistant infections, however resulting in some of the hindrances to antibiotic treatments [8, 9]. This review focused on host-directed therapies by using the immunomodulatory agents that target critical host signaling enzymes exploited by bacteria for their intracellular invasion, replication, and dissemination. We also describe the treatment of MDR Gram-negative bacterial infection with antibiotic-adjuvant combination and antibiotic-antibiotic combination.

MDR GRAM-NEGATIVE PATHOGENS (GNPS)

The number of Gram-negative bacterial infections is more than Gram-positive bacterial infections worldwide. AMR among Gram-negative bacteria is an emerging global problem because of the rapid spread of resistance mechanisms and restricted treatment regimens. Moreover, the incidence of serious infections from MDR-GNPs has increased dramatically and constituted a serious threat to world public health in the last decade [10 - 12]. The development of MDR strains (non-susceptible to >1 agent in >3 antimicrobial categories), extensively drug-resistant (XDR) strains (non-susceptible to >1 agent in all but <2 antimicrobial categories), and pan-drug-resistant (PDR) strains (non-susceptible to all antimicrobial agents), has increased abruptly. These infections have caused increasingly worsening morbidity and mortality and escalating treatment problems. Furthermore, they also have a great economic impact on healthcare environments due to the rising costs from prolonged hospitalizations. Moreover, the alarming aspect of these is that relatively few new antibiotics have been approved in recent years [13]. GNPs are the common causes of urinary tract infections (UTIs), intra-abdominal infections (IAIs), ventilator-associated pneumonia (VAP), and septicemia. Mainly, *Klebsiella pneumoniae, Escherichia coli*, and *Pseudomonas aeruginosa* are significant bacteria in the hospital infection, which account for 27% of all pathogens and 70% of all GNPs, causing healthcare-related infections [14]. The most common GNP causing VAP, and the second most common bacteria causing catheter-associated UTIs, is *P. aeruginosa*. The most common GNP causing central line-associated circulatory infections is *K. pneumoniae*. The most common GNPs causing UTIs and the second most common bacteria causing all of the healthcare-associated infections are *E. coli*. A majority of the mortalities are related to MDR-GNPs infections caused by carbapenem-resistant *Enterobacteriaceae* (CRE), extended-spectrum beta-lactamase (ESBL)-producing *Enterobacteriaceae*, MDR *P. aeruginosa*, and MDR *Acinetobacter baumannii* [15 - 18].

MECHANISM OF RESISTANT DRUGS

The expression of antibiotic inactivating enzymes and non-enzymatic pathways formulates the mechanism of AMR in GNP. The mechanisms take place by increasing the intrinsic resistance due to mutations in chromosomal genes or acquired mobile genetic elements carrying resistance genes. These include aminoglycosides modifying enzymes, plasmid encoding-lactamases, or non-enzymatic mechanisms like Qnr (plasmid-borne quinolone resistance gene) for fluoroquinolone (FQ) [19 - 21]. Notably, more than 10% of the *Enterobacteri-aceae* resisted 3rd generation cephalosporin, and approximately 2-7% of it resisted carbapenem. This resistance is caused by the fast distribution of extended-spectrum beta-lactamase (ESBL) producing strains Fig. (**1**). Additionally, the rates of carbapenem resistance for *K. pneumonia* are more than 25%. While *P. aeruginosa* and *A. baumannii* having resistance to carbapenem account for 20 to 40% and 40 to 70%, respectively [22].

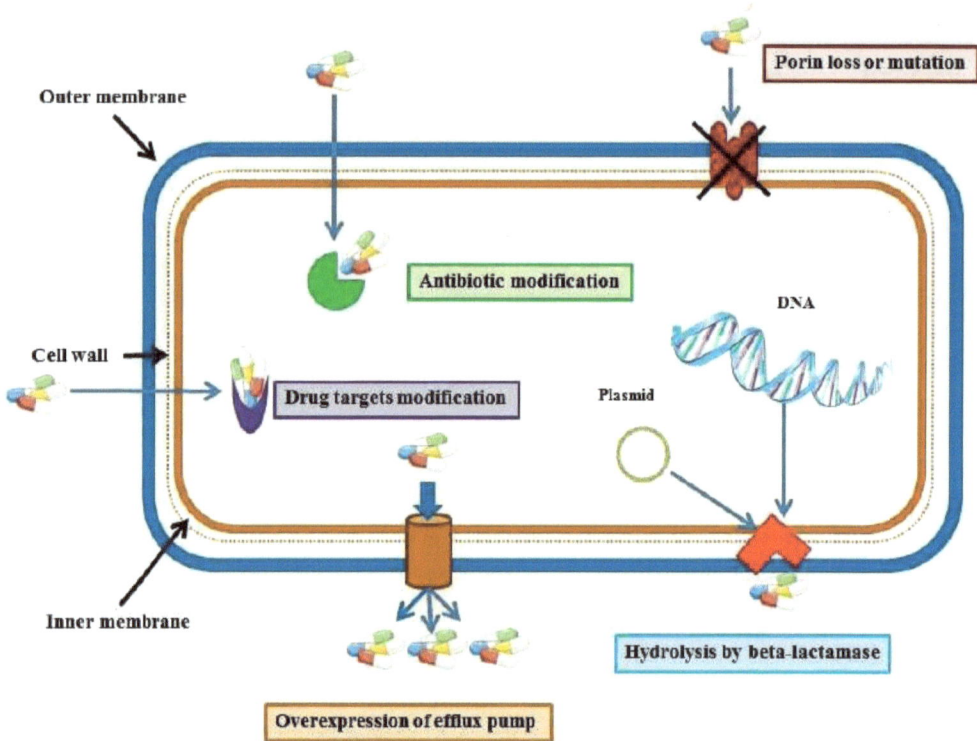

Fig. (1). Structure of Gram-negative bacteria and their mechanism of resistance.

ESBL-PRODUCING ENTEROBACTERIACEAE

Enterobacteriaceae family comprising *E. coli*, *Klebsiella* species, and *Enterobacter* species are the significant pathogens of UTIs, bloodstream infections, hospitals, and healthcare-related pneumonia. ESBLs production is the primary resistance mechanism, but other resistance mechanisms are also emerging, leading to MDR [23]. The principal mechanism of beta-lactam resistance in *Enterobacteriaceae* is beta-lactamase production [24]. *Enterobacteriaceae* and *Pseudomonas* species are currently a significant cause of hospital-acquired pneumonia (HAP), surgical wound infections, UTIs, and septicemias. Owing to the selection pressure of the antibiotic, these pathogens are highly efficient for the development of an acquiring gene that could resist many antibiotics. Two classifications are most commonly used for beta-lactamases: The Ambler scheme (molecular classification) and the Bush-Medeiros-Jacoby system (functional classification) (Table **1**) [25, 26]. The Ambler scheme divides beta-lactamases into class A, C, and D enzymes, which utilize serine for beta-lactam hydrolysis and class B metalloenzymes, which require divalent zinc ions for substrate hydrolysis. The Bush-Medeiros-Jacoby system includes group 1 cephalosporinases, group 2 broad-spectrum, inhibitor-resistant ESBLs, and serine carbapenemases, group 3 Metallo-beta-lactamases, and group 4 miscellaneous enzymes. The ESBLs are Ambler class A enzymes that hydrolyze extended-spectrum cephalosporin with an oxyimino side chain (ceftriaxone, cefotaxime, and ceftazidime). ESBLs are usually divided into 3 groups: TEM, SHV, and CTX-M [27]. In general, they are inhibited by beta-lactamase inhibitors, such as sulbactam, clavulanic acid, and tazobactam. As a plasmid harbors the Ambler Class A enzymes, they could easily be transmitted to different bacterial cells, causing rapid resistance to such enzymes. TEM-1 was first identified in 1965 in the *Enterobacteriaceae,* and it spread to *Neisseria, Haemophilus*, and *Vibrio* species. SHV-1 was found in 1979 and is usually detected in *Klebsiella* species and *E. coli* [28]. CTX-M was discovered by Tzouvelekis in 2000 [29]. Unlike TEM and SHV enzymes, CTX-M has no point mutation. There are 128 types of CTX-M reported and classified into CTX-M-1, CTX-M-2, CTX-M-8, CTX-M-9, and CTX-M-25 [30].

Table 1. The classification of beta-lactamases.

Ambler Scheme	Bush-Medeiros-Jacoby System	Characteristics of Beta-lactamases	No.
C	1	-Often chromosomal enzymes in Gram-negatives, but some are plasmid-coded -Not inhibited by clavulanic acid	51
A	2a	-Staphylococcal and enterococcal penicillinases	23
-	2b	-Broad-spectrum beta-lactamases including TEM-1 and SHV-1, mainly occurring in Gram-negatives	16
-	2be	-Extended spectrum beta-lactamases (ESBL)	200
-	2br	-Inhibitor-resistant TEM (IRT) beta-lactamases	24
-	2c	-Carbenicillin-hydrolyzing enzymes	19
-	2d	-Cloxacillin (oxacillin) hydrolyzing enzymes	31
-	2e	-Cephalosporinases inhibited by clavulanic acid	20
-	2f	-Carbapenem-hydrolyzing enzyme inhibited by clavulanic acid	4
B	3	-Metallo-enzymes that hydrolyze carbapenems and other beta-lactams except for monobactams -Not inhibited by clavulanic acid	24
D	4	-Miscellaneous enzymes that do not fit into other groups	9

CARBAPENEM-RESISTANT ENTEROBACTERIACEAE (CRE)

In 1990, CRE emerged in Japan and was followed by neighboring countries [31]. CRE produced Metallo-beta-lactamase (MBL) IMP-1, which hydrolyzed carbapenems and was encoded on plasmids that could transfer from one species to another. After that, another acquired MBL, VIM-1, was discovered. It was identified from *P. aeruginosa* and subsequently found in *Enterobacteriaceae* [32]. Furthermore, *K. pneumoniae* carbapenemase (KPC) was identified and found that it is encoded on a plasmid that could transfer and hydrolyze carbapenems oxyimino-cephalosporins efficiently [33]. In parallel to KPC's expansion, OXA-48 emerged and spread in *K. pneumoniae* in the Mediterranean countries [34]. Recently, a new group of MBL, NDM (New Delhi Metallo-beta-lactamase), was detected and demonstrated in carbapenem-resistant *E. coli* and *K. pneumoniae* in patients who had traveled from India [35]. After that, NDM-1 has spread extensively in South Asia and worldwide [36].

MDR *A. BAUMANNII*

The MDR pathogens commonly spread have been grouped in the acronym "ESKAPE," representing *Enterococcus faecium, Staphylococcus aureus, K. pneumoniae, A. baumannii, P. aeruginosa,* and *Enterobacter cloacae. A. baumannii* is one of the most severe ESKAPE pathogens that extensively resist antibacterial drugs [37]. In 1970, *A. baumannii* emerged as an important nosocomial pathogen related to the use of broad-spectrum antibiotics in the hospitals [38]. Recently, it has been found to resist most first-line antibiotics. Tigecycline and colistin are the only drugs that can be effective against it and are the last treatment choice for MDR *A. baumannii.* Moreover, colistin-resistant strains have been demonstrated in many countries worldwide [39]. Several *A. baumannii* strains are endemic to different regions worldwide; hence MDR *A. baumannii* resists antimicrobials through various mechanisms specific against particular types of antibiotics [40]. The armA gene encodes a 16S rRNA methylase that shows aminoglycoside resistance that was detected in *A. baumannii* in 2012 [41]. Moreover, MDR *A. baumannii* can show intrinsic chromosomal OXA-23, OXA-24, OXA-51-like, OXA-58-like, and OXA-12--like. The OXA-23 has been related to more outstanding production and dissemination of carbapenem resistance [42, 43]. MDR *A. baumannii* also resists fluoroquinolone by substituting the quinolone resistance-determining regions (QRDRs) of DNA topoisomerase IV and DNA gyrase, which interfere with the union of fluoroquinolone to the target proteins [44]. In 2005, Acinetobacter-derived cephalosporinases (ADC) beta-lactamase was reported in many clinical isolated specimens. This new enzyme was called ADC-7 beta-lactamase because six associated cephalosporinases had been described before [45]. Six years later,

Tian and colleagues reported the strains of *A. baumannii* that developed extended-spectrum class C beta-lactamases (ADC-33), which resisted against cefepime and other cephalosporins [46]. *A. baumannii* which resists sulbactam, is regarded for the reduction of penicillin-binding protein-2 (PBP2) expression [47]. Moreover, the gene blaTEM-1 also contributes to sulbactam resistance in *A. baumannii* [48]. Rifampicin is another antibiotic that *A. baumannii* resisted via rpoB mutations, active efflux, and enzymatic modification by the ADP ribosyltransferase of rifampicin (ARR-2) [49]. The resistance against tetracyclines is mediated by the active efflux of the antimicrobial drugs and the inhibition of ribosomal and tetracycline binding [50]. For polymyxin resistance by *A. baumannii*, there are two mechanisms for developing drug resistance. These are modifying lipopolysaccharides (LPS) lipid A and complete loss of the initial LPS [51, 52].

HOST-DIRECTED THERAPIES

With the growing problem of MDR-GNPs and diminishing antibiotic treatment options, the search for new treatment choices is now focused on generating effective clinical and immunological responses using host-directed therapies (HDTs) [53, 54]. HDTs target host-encoded functions necessary for bacterial infection, replication, virulence, and pathogenesis, which can be an alternative strategy to conventional antibiotics. In contrast to antimicrobial drugs that directly act on bacteria, HDTs intensify a potentially broad response against the pathogens, modifying them to be less prone to microbial resistance. Bacteria can combat the body's natural defense systems by several methods to escape recognition by the host's immune system. Some bacteria are obligate intracellular pathogens that reside in the host cell environment, especially within acidic vesicles. These bacteria must take over the host cell mechanisms. The extensive knowledge of bacteria's activities to influence crucial machinery inside the host cells has inspired the development of HDTs. Most of these drugs are novel small molecules or peptides approved by the FDA for other indications [54]. Moreover, emerging research indicates that HDTs might be the most effective treatment when co-administered with conventional antibiotics and could reduce the antibiotic dosage necessary to treat MDR bacteria [55, 56].

In 2017, Andersson and colleagues reported that 121 non-antibiotic drugs were demonstrated to inhibit bacterium-induced cytotoxicity in murine macrophages. The macrophage cytotoxicity induced by two additional MDR strains was inhibited by 13 of 121 non-antibiotic drugs. The intracellular survival of *A. baumannii* in macrophages was decreased in 6 of 121 non-antibiotic drugs [57]. In 2018, Jatana and colleagues reported the efficacy of a pyrimidine synthesis inhibitor, N-phospho-acetyl-L-aspartate (PALA), to increase MDR's clearance *A. baumannii* strains by primary human dermal fibroblasts *in vitro*. The PALA did not directly inhibit the pathogen but increased cellular secretion of the antimicrobial peptides [human β-defensin 2 (HBD2)] and HBD3 fibroblasts. These results indicate that PALA might be a new option to treat MDR *A. baumannii* infections of the skin by enhancing the cutaneous innate immune defense system [58]. Recently, Guo *et al.* induced *A. baumannii* infection in mice with chronic pulmonary inflammation after intranasal administration of SiO_2 and found that SiO_2 treatment could increase host defense against *A. baumannii*. Innate immune responses initiated by NF-κB, type 1 interferon, NLRP3, and AIM2 inflammasomes were dispensable for SiO_2-mediated host defense. SiO_2 treatment also activated the mTORC1 signaling, which was crucial for host defense against *A. baumannii* infection [59]. In 2016, Abdulnour and colleagues reported that aspirin-triggered resolvin D1 (AT-RvD1) could increase efferocytosis in murine lung macrophages from *E. coli*. Early treatment with exogenous AT-RvD1 enhanced clearance of *E. coli* and *P. aeruginosa in vivo*. They concluded that these antibacterial and pro-resolving actions of AT-RvD1 were additive to antibiotic therapy as HDT in MDR *E. coli* and *P. aeruginosa* infection [60].

ANTIBIOTIC-ADJUVANT COMBINATION THERAPY

Gram-negative bacterial resistance is one of the most common problems that impact human health. The antibiotic-adjuvant combination can increase the antibacterial activity and prolong the life span of presently used antibiotics by directly inhibiting the resistance or complementary mechanism, which provides an alternative strategy to fight MDR Gram-negative bacteria. Contrary to the development of new antibiotics, the adjuvants of broad-spectrum antibiotics would be more hopeful and cost-effective for the modern healthcare environment. The hybridization technique can maximize the efficacy of adjuvants and avoid the un-coincident pharmacological properties of 2 drugs *in vivo*.

In 2012, Chauhan and colleagues reported ethylene diamine tetraacetic acid (EDTA) combined with gentamicin to eliminate mature biofilms using an *in vivo* study of an implantable venous access port which was inserted in rats and colonized by *E. coli* or *P. aeruginosa*. They found that 30 mg/ml EDTA combined with 5 mg/ml gentamicin could completely eradicate Gram-negative bacterial biofilms in implantable venous access ports. They concluded that EDTA was an effective antibiotic adjuvant to eliminate biofilms of Gram-negative bacterial pathogens on the catheter and provide a promising new lock solution [61]. Yenn and colleagues conducted the study to evaluate stigmasterol's synergistic antibiotic effect as an adjuvant of ampicillin for beta-lactamas--producing clinical isolates. They found that this combination demonstrated importantly better antibiotic action on all tested bacteria. This combination greatly decreased the colony counts by 98.7%. They concluded that stigmasterol showed the potential of antibiotic adjuvant and could increase the antibiotics' efficacy [62]. Stigmasterol is an unsaturated phytosterol from the plant fats or oils of many kinds of plants. It is used as a food additive in manufactured food products in the United Kingdom and the European Union [63]. In 2017, Zurawski and colleagues evaluated the combination of SPR741, which permeabilizes the Gram-negative bacterial membrane, and rifampin against XDR and AB5075 *A. baumannii*. They conducted an *in vitro* study in a murine pulmonary model and found that this drug combination could significantly decrease bacterial infection and improve animal survival from an extensive infection [64]. SPR741, a polymyxin-B-derived molecule, was specially developed to minimize the nephrotoxicity related to this class of antibiotics. SPR741 could reduce the positive charge and decreased the fatty-acid side chain with a highly lipophilic presence of polymyxin. These two structural features were responsible for clinical nephrotoxicity [65]. In the same year, Durand-Réville and colleagues reported the most important combinations of sulbactam and ETX2514, which showed potent antibacterial action against MDR *A. baumannii* infections. The ETX2514 is an avibactam-derived expanded-spectrum serine beta-lactamase inhibitor, which demonstrated highly potent inhibiting action to clinically relevant class A, C, and D beta-lactamases and penicillin-binding proteins (PBPs). It could restore beta-lactam activity against a broad range of MDR Gram-negative pathogens. They concluded that sulbactam–ETX2514 might significantly benefit MDR A. baumannii infected patients and the global healthcare system [66]. In 2019, Baker and colleagues conducted a study to evaluate the effects of the combination between the peptides and azithromycin or rifampicin against MDR Gram-negative bacteria. They found that two peptides (KLWKKWKKWLK-NH2 and GKWKKILGKLIR-NH2) showed a synergistic effect of azithromycin and clindamycin in MDR *E. coli* ST131 and *K. pneumoniae* ST258. Moreover, the low concentration of the 2 peptides combined with azithromycin and rifampicin could inhibit *E. coli, K. pneumoniae*, and *A. baumannii* strains. They concluded that the intrinsic resistance to azithromycin and rifampicin in Gram-negative bacteria could be overcome by low peptide concentrations, with no toxicity to eukaryotic cells [67]. Recently, Song and colleagues reported the boosting effect of all major classes of antibiotics, including cefepime, ofloxacin, colistin, tetracycline, rifampicin, and vancomycin, against MDR Gram-negative pathogens using the short linear antibacterial peptide (SLAP)-S25, which carried four non-natural amino acids of 2,4-diaminobutanoic acid (Dab) and showed weak antibacterial activity. They found that SLAP-S25 could trigger membrane damage by binding to lipopolysaccharide (LPS) in the outer membrane and phosphatidylglycerol (PG) in the bacterial cytoplasmic membrane, potentiaing antibiotic efficacy through combined effects. Furthermore, SLAP-S25 effectively increased the action of colistin against MDR *E. coli*-associated infections in animal models. They concluded that a potential therapeutic strategy using existing antibacterial drugs combined with SLAP-S25 could address the prevalent infections caused by MDR Gram-negative pathogens worldwide [68]. Douafer and colleagues developed the first combination of adjuvant and antibiotic in an aerosolized form and demonstrated this combination's feasibility. The combination aerosol droplets have been designed in sizes suitable for inhalation between 3.4 and 4.4 μm as measured for the adjuvant polyamino-isoprenyl compound (NV716) and doxycycline, respectively. The results suggested that the doxycycline/NV716 combination could be successfully delivered at the required concentration in the lungs and could reduce drug utilization and increase treatment efficacy. They concluded that the doxycycline/NV716 combination showed a potentially important drug for treating *P. aeruginosa* pulmonary infections. Moreover, the novel pharmaceutical development strategies could supply an adjunct pulmonary treatment to reduce the therapeutic duration and improve the current drug protocols [69]. Recently, Barker and colleagues identified 2 natural products which showed a synergistic effect with colistin. The first natural product was ascomycin, the non-antimicrobial macrolide produced by the fermentation of *Streptomyces hygroscopicus*, that constantly potentiated colistin in several Gram-negative strains. The second one was kuwanon G, which was isolated from *Morus alba* and demonstrated consistent action in several highly colistin-resistant clinical isolates, and restored colistin MICs. The results showed that both ascomycin and kuwanon G exerted dose-dependent action and had no effects on bacterial growth in monotherapy. Colistin is a cyclic, poly-cationic antimicrobial drug used as a last antimicrobial drug for treating MDR Gram-negative bacterial infections. Cell signaling proteins involved in colistin resistance mechanisms display both kinase and phosphatase enzymes. They found that kuwanon G could break colistin resistance, while ascomycin could potentiate colistin in polymyxin susceptible bacteria [70].

ANTIBIOTIC-ANTIBIOTIC COMBINATION THERAPY

In 2019, Schmid and colleagues conducted a systematic review and meta-analysis related to monotherapy vs. combination therapy for MDR Gram-negative infections. They found that the combination of antimicrobial therapy for MDR Gram-negative bacteria appears to be superior to monotherapy concerning mortality [71]. There are no RCTs to compare the combination therapy with monotherapy for carbapenem resistance *Enterobacteriaceae* (CRE) infections. The CRE is predominantly attributed to beta-lactamase production. The common pathogens which produced carbapenemases encountered in clinical practice are *K. pneumoniae*, *P. aeruginosa*, and *A. baumannii* [72, 73].

TREATMENT OF MDR *K. PNEUMONIAE*

Carbapenems resistance in *K. pneumoniae* involves multiple mechanisms that include an alteration in outer membrane permeability mediated by the loss of porins, carbapenemases (KPC, NDM, VIM, OXA-48 like), and the up-regulation of efflux systems [74]. *K. pneumoniae* with carbapenemases infection is usually related to high mortality rates attributed to the insufficiency of active antimicrobial drugs and underlying comorbidities of the patients [75]. The antibiotic-antibiotic combination therapy involves heterogeneity of double- and triple-combination regimens. The most commonly administered antimicrobial drugs in double or triple combinations are carbapenem plus colistin, colistin plus tigecycline, carbapenem plus carbapenem, and carbapenem plus colistin plus tigecycline [76 - 78]. In 2018, Abdelsalam and colleagues conducted a prospective and comparative study to evaluate the effectiveness and adverse effects of colistin alone [a loading dose of 9 million international units (MIU) followed by 3 MIU every 8 h] *vs.* colistin plus meropenem 1 gram every 8 h in 60 MDR *K. pneumoniae*-induced HAP or VAP patients. They found that colistin-meropenem combination treatment showed a significant decrease in mortality *vs.* colistin alone [16.7% (5/30) *vs.* 43.3% (13/30); p = 0.047]. Furthermore, the combination therapy was not associated with hepatotoxicity, nephrotoxicity, or neurotoxicity. They concluded that the colistin-meropenem combination therapy was better than the colistin monotherapy in treating MDR *K. pneumoniae*-induced HAP or VAP [79]. In the same year, Paul and colleagues conducted a randomized controlled superiority trial in six hospitals in Greece, Israel, and Italy. They randomly assigned 406 pneumonia or bacteremia patients to the two treatment groups [intravenous colistin (9-million-unit loading dose, followed by 4.5 million units twice per day) or colistin with meropenem (2-g prolonged infusion three times per day)]. Most infections were caused by *A. baumannii* (77%). They found no significant difference between colistin monotherapy and combination therapy for clinical failure 14 days after randomization. They concluded that this combination therapy was not superior to monotherapy. Moreover, the combination of meropenem with colistin did not improve clinical outcomes in severe *A. baumannii* infections. However, when focusing on carbapenem-resistant *Enterobacteriaceae*, specifically with *K. pneumoniae*, a trend towards lower clinical failure and mortality rates was observed [80]. After that, Wunderink *et al.* conducted a phase 3, multinational, open-label RCT (TANGO II) to evaluate the efficacy and safety of meropenem-vaborbactam therapy vs. best available therapy (BAT) for CRE. Seventy-seven patients with confirmed or suspected CRE infection were randomized 2:1 to meropenem-vaborbactam (2 g/2 g over 3 h, q8h for 7-14 days) or BAT. They found that treatment with meropenem-vaborbactam resulted in a significantly higher clinical outcome rate and test of cure at the end of treatment. A trend towards lower mortality rates and slightly lower rates of nephrotoxicity in the intervention group could be demonstrated, although not statistically significant [81]. In 2015, Ji and colleagues conducted a trial to test the antimicrobial activities of a combination of cefepime, amoxicillin, and clavulanic acid *vs.* tigecycline in treating hospital-acquired blaKPC-positive *K. pneumoniae* patients. The pathogens demonstrated high-grade resistance to cephalosporins and carbapenems. Combining cefepime with amoxicillin and clavulanic acid significantly reduced the MIC against respective pathogens *in vitro*. Twenty-six patients were enrolled in the study group, and 25 cases were enrolled in the control groups. The overall pathogen clearance rate and the mortality showed no significant differences ($p = 0.447$ and $p = 0.311$). The total cost and the portion of the cost not covered by insurance were higher for the control group than the study group ($p < 0.001$). They concluded that cefepime plus amoxicillin/clavulanic acid combination was an effective and economical option for KPC-*K. pneumoniae* infection [82]. Another observational, case-control (1:2) study involving critically ill adult patients with a microbiologically confirmed CR-*K. pneumoniae* invasive infection treated with the double carbapenem (DC) regimen matched with those receiving standard treatment was conducted by De Pascale *et al.* in 2017. The reason was that in severe infection patients with existing contraindications for gentamicin or colistin treatments, the treatment with ertapenem might comparatively inhibit produced carbapenemases and enhanced activity of the added second carbapenem. Forty-eight patients treated with carbapenem plus carbapenem were matched with 96 controls. On day 28, they found a significant reduction in mortality rate, meaning that carbapenem-carbapenem combination treatment had a survival benefit compared with standard treatment (29.2% vs. 47.9%, $p = 0.04$). Moreover, microbiological eradication and clinical cure were significantly higher in the carbapenem-carbapenem combination as compared to the standard treatment (57.9% vs. 25.9%, $p = 0.04$, and 65% *vs.* 31.3%, $p = 0.03$, respectively). They concluded that an improved 28-day mortality rate was associated with a carbapenem-carbapenem combination regimen compared with the standard treatment for severe CR-*K. pneumoniae* infections [83]. In summary, the antibiotic combination therapy demonstrated a higher rate of clinical cure-associated outcomes and reduced the mortality rate [84].

TREATMENT OF MDR *P. AERUGINOSA*

P. aeruginosa develops almost all the resistance mechanisms to protect itself from the antibiotic activity and is one of the most challenging pathogens to be treated. The most crucial resistance mechanism involves the efflux pumps that could eliminate beta-lactams, fluoroquinolones, many dyes and detergents. Other mechanisms include the loss of impermeability due to loss of porin OprD, the production of inducible AmpC enzyme, which is mainly associated with reduced susceptibility to carbapenems, and acquired enzymatic mechanisms of resistance, such as OXA, ESBL enzymes, and MBL enzymes, which spread of MDR [85]. Ceftolozane/tazobactam (C/T) is a novel beta-lactam/beta-lactamase inhibitor (BL/BLI), which demonstrated excellent *in vitro* activity against MDR *P. aeruginosa*. US FDA approved this new drug for the treatment of complicated intra-abdominal infections (cIAI) and complicated urinary tract infections (cUTI) due to the strength of the results of the ASPECT-cIAI and ASPECT-cUTI trials [86, 87]. Recently, C/T has also been approved for HAP and VAP since high-dose C/T (3 g q 8 h) achieved noninferiority *vs.* meropenem in ASPECT-nosocomial pneumonia (NP) trial [88]. In 2019, Bassetti and colleagues conducted a retrospective study that included adult patients treated with ≥ 4 days of C/T from 22 hospitals in Italy. C/T treatment was documented in 101 patients with various infections, including NP (31.7%), acute bacterial skin and skin structure infection (20.8%), cUTI (13.9%), cIAI (12.9%), bone infection (8.9%), and primary bacteremia (5.9%). Over one-half of *P. aeruginosa* strains were XDR (50.5%), with 78.2% isolates resistant to at least one carbapenem. In this study, C/T was used as first-line therapy in 39 patients (38.6%), and concomitant antibiotics were reported in 35.6% of the patients. C/T doses were 1.5 g q8h in 70 patients (69.3%) and 3 g q8h in 31 patients (30.7%); the median duration of C/T therapy was 14 days. Overall clinical success was 83.2%. They concluded that C/T demonstrated a safety and tolerability profile regardless of the infection type. Clinicians should be aware of the risk of clinical failure with C/T therapy in septicemia patients receiving continuous renal replacement therapy [89]. Moreover, Stone and colleagues evaluated ceftazidime/avibactam (C/A) against MDR *Enterobacteriaceae* and *P. aeruginosa* isolates pooled from the adult phase 3 clinical trials in patients with cIAI, cUTI, or NP, including VAP. They included isolates from 1,051 patients with MDR *Enterobacteriaceae* and 95 patients with MDR *P. aeruginosa*. They found that the favorable microbiological response rates at test-of-cure (TOC) for all MDR *Enterobacteriaceae* and MDR *P. aeruginosa* were 78.4% and 57.1% C/A and 71.6% and 53.8%, respectively, for comparators. The proportions of patients with ≥1 MDR isolate who were clinically cured at TOC were similar in the C/A arms (85.4%) and comparator (87.9%) arms. They concluded that C/A demonstrated the same clinical efficacy to predominantly carbapenem comparators against MDR *Enterobacteriaceae* and *P. aeruginosa* and maybe a suitable regimen to carbapenem-based therapies cIAI cUTI, and NP/VAP caused by MDR Gram-negative pathogens [90]. In 2018, Torres and colleagues conducted the study to evaluate the efficacy and safety of C/A in patients with nosocomial pneumonia compared with meropenem in a multinational, phase 3, double-blind, noninferiority trial (REPROVE). They randomly assigned 879 patients from 136 centers in 23 countries to 2,000 mg ceftazidime and 500 mg avibactam as 1:1 ration (by 2 h intravenous infusion every 8 h) or 1,000 mg meropenem (by 30-min intravenous infusion every 8 h) for 7-14 days. They included 808 patients in the safety population, 726 patients in the clinically modified intention-to-treat population, and 527 patients in the clinically evaluable population. The Gram-negative baseline pathogens in the microbiologically modified intention-to-treat population were *K. pneumoniae* (37%) and *P. aeruginosa* (30%). Two hundred and forty-five of 356 patients (68.8%) in the C/A group were clinically cured compared with 270 of 370 patients (73.0%) in the meropenem group in the clinically modified intention-to-treat population. One hundred ninety-nine of 257 participants (77.4%) were clinically cured in the C/A group, compared with 211 of 270 patients (78.1%) in the meropenem group, in the clinically evaluable population. They concluded that C/A was non-inferior to meropenem in the treatment of nosocomial pneumonia [91].

If a combination of empiric therapy is used, two drugs from different classes with *in vitro* activity against *P. aeruginosa* are recommended. Commonly, beta-lactam is used in combination with polymyxin, aminoglycoside, or fluoroquinolone. Aminoglycosides (amikacin and tobramycin) demonstrated reliable action against many MDR-*P. aeruginosa* and maybe the only active agent for XDR strains. Similarly, polymyxins demonstrated *in vitro* activity against many MDR and XDR strains [92].

TREATMENT OF MDR *A. BAUMANNII*

The *A. baumannii* usually develop multiple resistance mechanisms to many antibiotics classes and have become difficult-to-treat resistance (DTR) organisms in healthcare-associated infections. The most important mechanisms of resistance are the production of chromosomally encoded AmpC cephalosporinases, the production of carbapenemases, the changes in the porins, the production of aminoglycoside-modifying enzymes, and the efflux pumps [93]. In 2013, Aydemir and colleagues conducted the RCT to compare the responses of colistin treatment alone vs. a combination of rifampicin and colistin in VAP treatment caused by a carbapenem-resistant *A. baumannii* strain. They randomly assigned 43 patients to one of two treatment groups. They found that clinical, laboratory, radiological, and microbiological response rates were better in the combination. However, time to microbiological clearance was significantly shorter in the combination group. The VAP-related mortality rates were 63.6% and 38.1% for the colistin and the combination groups. They concluded that the combination of rifampicin with colistin might improve the clinical and microbiological outcomes of VAP patients infected with MDR *A. baumannii* [94]. In the same year, Durante-Mangoni and colleagues conducted the multicenter, parallel, randomized, open-label clinical trial to enroll 210 patients with life-threatening infections due to XDR *A. baumannii* from ICUs of 5 tertiary care hospitals to treat with colistin alone or colistin plus rifampicin. They found that no difference was observed for infection-related death and length of hospitalization. However, a significant increase in the microbiologic clearance rates was observed in the colistin plus rifampicin arm. They concluded that rifampicin should not be routinely combined with colistin in clinical practice [95]. In 2015, Sirijatuphat *et al.* conducted the RCT to enroll 44 patients infected with carbapenem-resistant *A. baumannii* to treat with colistin alone or colistin plus fosfomycin for 7 to 14 days. They found that the patients treated with combination therapy had a significantly more favorable microbiological response and a trend toward more favorable clinical outcomes and lower mortality than those who received colistin alone [96]. In 2018, Qin and colleagues conducted the RCT to study the effect of high-dose cefoperazone-sulbactam combined with tigecycline in VAP patients infected by XDR *A. baumannii*. Forty-two patients with VAP were randomized into the tigecycline group and the tigecycline plus cefoperazone-sulbactam group. They found that the total combined effectiveness rate was significantly higher in the combination group (85.7%) compared with the tigecycline group (47.6%) (p = 0.010). No significant differences were noted concerning the side effects between the two groups. They concluded that the high dose cefoperazone-sulbactam could improve tigecycline antimicrobial activity against XDR *A. baumannii* infection [97]. In the same year, Mosaed and colleagues conducted a single-blinded RCT to enroll 29 adult patients with VAP infected with MDR *A. baumannii* to receive colistin plus levofloxacin or ampicillin-sulbactam plus levofloxacin. They found that a clinical response occurred in 3 (27%) and 10 (83%) in colistin plus levofloxacin and ampicillin-sulbactam plus levofloxacin arms, respectively (p = 0.007). They concluded that levofloxacin plus high dose ampicillin-sulbactam is more effective than levofloxacin plus colistin in the VAP patients infected with MDR *A. baumannii* with a significantly lower risk of nephrotoxicity [98]. After that, Khalili and colleagues conducted the RCT to evaluate the efficacy of colistin plus meropenem compared with ampicillin-sulbactam plus meropenem in VAP patients' treatment infected with carbapenem-resistant *A. baumannii*. Forty-seven VAP patients were randomized to receive meropenem/colistin or meropenem/ampicillin-sulbactam for 14 days. They found that the clinical response (75% *vs.* 69.6%, p = 0.75) and microbial eradication (87.50% *vs.* 91.3%, p = 0.59) were no different between meropenem/colistin and meropenem/-ampicillin-sulbactam groups, respectively. They concluded that the clinical and microbiological responses were comparable between the meropenem/colistin and meropenem/ampicillin-sulbactam groups [99]. In 2019, Pourheidar and colleagues conducted a randomized, open-label trial to evaluate the efficacy and safety of ampicillin-sulbactam plus nebulized colistin compared with colistin plus nebulized colistin in the treatment of VAP caused by MDR *A. baumannii* in ICU patients. They found no significant difference between the 2 groups in the rate of microbiological eradication, clinical outcome, survival rate, and length of stay in hospital or ICU. However, the comparison of cumulative patient-days with stages 2 and 3 AKI revealed a significant difference in colistin plus nebulized colistin group *vs.* ampicillin-sulbactam plus nebulized colistin group (p = 0.013). They concluded that the high dose ampicillin-sulbactam plus nebulized colistin regimen has comparable efficacy with colistin plus nebulized colistin in VAP treatment caused by MDR *A. baumannii* with the lower incidence of kidney injury [100].

TREATMENT OF MDR *E. COLI*

In 2018, Harris and colleagues conducted the noninferiority, parallel-group RCT to evaluate the efficacy of piperacillin-tazobactam compared with meropenem in patients with bloodstream infection caused by ceftriaxone resistant *E. coli*. One hundred and eighty-eight patients were randomly assigned to piperacillin-tazobactam, and 191 patients were randomly assigned to meropenem for a minimum of 4 - 14 days determined by the physician. They found that 23 of 187 patients (12.3%) in the piperacillin-tazobactam group met the primary outcome of mortality at 30 days compared with 7 of 191 (3.7%) in meropenem. They concluded that among patients with *E. coli* bloodstream infection and ceftriaxone resistance, the treatment with piperacillin-tazobactam compared with meropenem did not result in non-inferior 30-day mortality [101]. In 2017, Sims and colleagues conducted the prospective, randomized, double-blind, Phase 2 dose-ranging trial to compare the efficacy and safety of imipenem/cilastatin plus relebactam 125 mg or 250 mg with imipenem/cilastatin alone in patients with cUTIs (1:1:1). They recruited 300 cUTI or acute pyelonephritis patients who needed hospitalization and IV antibacterial therapy. The primary endpoint was a favorable microbiological response rate (pathogen eradication) at discontinuation of intravenous therapy (DCIV) in the microbiologically evaluable (ME) population. They found that at DCIV, 71 patients in the imipenem/cilastatin plus relebactam 250 mg, 79 patients in the imipenem/cilastatin plus relebactam 125 mg, and 80 patients in the imipenem/cilastatin alone group were ME. The microbiological response rates were 95.5%, 98.6%, and 98.7%, respectively, confirming noninferiority of both imipenem/cilastatin plus relebactam doses to imipenem/cilastatin alone. The clinical response rates were 97.1%, 98.7%, and 98.8%, respectively. They concluded that imipenem/cilastatin plus relebactam (250 or 125 mg) was as effective as imipenem/cilastatin alone to treat cUTI patients infected with MDR *E. coli* [102]. In 2018, Kaye and colleagues conducted a phase 3, multicenter, multinational RCT (TANGO I) to compare meropenem-vaborbactam piperacillin-tazobactam adults with cUTIs or acute pyelonephritis. The primary efficacy endpoint was evaluated using the FDA and European Medicines Agency (EMA) criteria. They randomized a total of 550 adults with cUTIs to the treatment groups. Five hundred and forty-five patients received at least 1 dose of the study drugs (272 received meropenem-vaborbactam and 273 received piperacillin-tazobactam), and 374 (68%) comprised the microbiologically modified intent-to-treat population of the randomized patients. The most common urinary pathogens in the trial were *E. coli*. They found that the noninferiority was met for the FDA and EMA primary endpoints. For the FDA primary endpoint, at the end of treatment, the overall success rate in the microbiologically modified intent-to-treat population was 98% with meropenem-vaborbactam *vs.* 94% with piperacillin-tazobactam. For the EMA primary endpoint, microbial eradication in the microbiologically modified intent-to-treat population occurred in 67% of the meropenem-vaborbactam group vs. 58% piperacillin-tazobactam group (p < 0.001 for noninferiority). They concluded that among patients with cUTIs, meropenem-vaborbactam *vs.* piperacillin-tazobactam demonstrated a composite outcome of complete resolution or improvement of symptoms along with microbial eradication that met the noninferiority criterion [103]. From the TANGO II study conducted by Wunderink and colleagues, they found that meropenem-vaborbactam for CRE *E. coli* infection was associated with the increased clinical cure, decreased mortality, and reduced nephrotoxicity compared with BAT [81].

CONCLUDING REMARKS

Regardless of increased efforts to develop new antibiotics, the number of drug approvals has diminished in the last few years. Since 2010, less than 10 new drugs with activity against MDR Gram-negative bacteria have been approved by the FDA and the EMA, as ceftolozane/tazobactam, ceftazidime/avibactam, meropenem/vaborbactam, plazomicin, and eravacycline. To protect these new antibiotics from resistance development, it is critical to use them only as the most effective treatments when infections caused by Gram-negative bacteria resistant to the possible options are confirmed or strictly suspected. Although there are various new antibiotics currently in clinical trials, most of them are mostly modifications of existing antibiotic classes addressing specific resistant mechanisms and are active only against specific pathogens or limited subsets of resistant strains. On the other hand, new alternative treatments, such as host-directed, antibiotic-adjuvant combination, and antibiotic-antibiotic combination therapy, are in development to manage MDR Gram-negative bacterial infections, and their impact is yet to be explored. Finally, colistin is still considered as a fundamental antibiotic for the treatment of CRE, CR-*P. aeruginosa* and CR-*A. baumannii* in combination with suitable antibiotics. Moreover, ceftazidime-avibactam and meropenem-vaborbactam can be used as the backbones in the treatment of CRE. Similarly, ceftolozane-tazobactam could be used as a beta-lactam backbone for the treatment of MDR *P. aeruginosa*. However , the novel combination of antibiotics is not the panacea for the continuing crisis in MDR Gram-negative bacteria therapy.

CONSENT FOR PUBLICATION

Not applicable.

CONFLICT OF INTEREST

The author declares no conflict of interest, financial or otherwise.

ACKNOWLEDGEMENTS

Declared none.

REFERENCES

[1]   Cassini A, Högberg LD, Plachouras D, *et al.* Burden of AMR Collaborative Group. Attributable deaths and disability-adjusted life-years caused by infections with antibiotic-resistant bacteria in the EU and the European Economic Area in 2015: a population-level modelling analysis. Lancet Infect Dis 2019; 19(1): 56-66.
[http://dx.doi.org/10.1016/S1473-3099(18)30605-4] [PMID: 30409683]

[2]   Tamma PD, Aitken SL, Bonomo RA, Mathers AJ, van Duin D, Clancy CJ. Infectious diseases society of america antimicrobial resistant treatment guidance: gram-negative bacterial infections. Clin Infect Dis 2020; ciaa1478.
[http://dx.doi.org/10.1093/cid/ciaa1478]

[3]   Tornimbene B, Eremin S, Escher M, Griskeviciene J, Manglani S, Pessoa-Silva CL. WHO global antimicrobial resistance surveillance system early implementation 2016-17. Lancet Infect Dis 2018; 18(3): 241-2.
[http://dx.doi.org/10.1016/S1473-3099(18)30060-4] [PMID: 29396007]

[4]   Theuretzbacher U. Global antimicrobial resistance in Gram-negative pathogens and clinical need. Curr Opin Microbiol 2017; 39: 106-12.
[http://dx.doi.org/10.1016/j.mib.2017.10.028] [PMID: 29154024]

[5]   Magiorakos AP, Srinivasan A, Carey RB, *et al.* Multidrug-resistant, extensively drug-resistant and pandrug-resistant bacteria: an international expert proposal for interim standard definitions for acquired resistance. Clin Microbiol Infect 2012; 18(3): 268-81.
[http://dx.doi.org/10.1111/j.1469-0691.2011.03570.x] [PMID: 21793988]

[6]   Hancock RE, Nijnik A, Philpott DJ. Modulating immunity as a therapy for bacterial infections. Nat Rev Microbiol 2012; 10(4): 243-54.
[http://dx.doi.org/10.1038/nrmicro2745] [PMID: 22421877]

[7]   Tamma PD, Cosgrove SE, Maragakis LL. Combination therapy for treatment of infections with gram-negative bacteria. Clin Microbiol Rev 2012; 25(3): 450-70.
[http://dx.doi.org/10.1128/CMR.05041-11] [PMID: 22763634]

[8]   Kubelkova K, Macela A. Innate immune recognition: an issue more complex than expected. Front Cell Infect Microbiol 2019; 9: 241.
[http://dx.doi.org/10.3389/fcimb.2019.00241] [PMID: 31334134]

[9]   Ulevitch RJ, Tobias PS. Recognition of gram-negative bacteria and endotoxin by the innate immune system. Curr Opin Immunol 1999; 11(1): 19-22.
[http://dx.doi.org/10.1016/S0952-7915(99)80004-1] [PMID: 10047547]

[10]   Xu ZQ, Flavin MT, Flavin J. Combating multidrug-resistant Gram-negative bacterial infections. Expert Opin Investig Drugs 2014; 23(2): 163-82.
[http://dx.doi.org/10.1517/13543784.2014.848853] [PMID: 24215473]

[11]   Kunz AN, Brook I. Emerging resistant Gram-negative aerobic bacilli in hospital-acquired infections. Chemotherapy 2010; 56(6): 492-500.
[http://dx.doi.org/10.1159/000321018] [PMID: 21099222]

[12]   Kollef MH, Golan Y, Micek ST, Shorr AF, Restrepo MI. Appraising contemporary strategies to combat multidrug resistant gram-negative bacterial infections--proceedings and data from the Gram-Negative Resistance Summit. Clin Infect Dis 2011; 53 (Suppl. 2): S33-55.
[http://dx.doi.org/10.1093/cid/cir475] [PMID: 21868447]

[13]   Sievert DM, Ricks P, Edwards JR, *et al.* National Healthcare Safety Network (NHSN) Team and Participating NHSN Facilities. Antimicrobial-resistant pathogens associated with healthcare-associated infections: summary of data reported to the National Healthcare Safety Network at the Centers for Disease Control and Prevention, 2009-2010. Infect Control Hosp Epidemiol 2013; 34(1): 1-14.
[http://dx.doi.org/10.1086/668770] [PMID: 23221186]

[14]   Pfeifer Y, Cullik A, Witte W. Resistance to cephalosporins and carbapenems in Gram-negative bacterial pathogens. Int J Med Microbiol 2010; 300(6): 371-9.
[http://dx.doi.org/10.1016/j.ijmm.2010.04.005] [PMID: 20537585]

[15]   Perez F, Van Duin D. Carbapenem-resistant *Enterobacteriaceae*: a menace to our most vulnerable patients. Cleve Clin J Med 2013; 80(4): 225-33.
[http://dx.doi.org/10.3949/ccjm.80a.12182] [PMID: 23547093]

[16]   Vink J, Edgeworth J, Bailey SL. Acquisition of MDR-GNB in hospital settings: a systematic review and meta-analysis focusing on ESBL-E. J Hosp Infect 2020; 106(3): 419-28.
[http://dx.doi.org/10.1016/j.jhin.2020.09.006] [PMID: 32918969]

[17]   Snitkin ES, Zelazny AM, Thomas PJ, *et al.* NISC Comparative Sequencing Program Group. Tracking a hospital outbreak of carbapenem-resistant *Klebsiella pneumoniae* with whole-genome sequencing. Sci Transl Med 2012; 4(148): 148ra116.
[http://dx.doi.org/10.1126/scitranslmed.3004129] [PMID: 22914622]

[18]   Zgurskaya HI, Löpez CA, Gnanakaran S. Permeability barrier of Gram-negative cell envelopes and approaches to bypass it. ACS Infect Dis 2015; 1(11): 512-22.
[http://dx.doi.org/10.1021/acsinfecdis.5b00097] [PMID: 26925460]

[19]   Nikaido H. Molecular basis of bacterial outer membrane permeability revisited. Microbiol Mol Biol Rev 2003; 67(4): 593-656.
[http://dx.doi.org/10.1128/MMBR.67.4.593-656.2003] [PMID: 14665678]

[20]   Fernández L, Hancock RE. Adaptive and mutational resistance: role of porins and efflux pumps in drug resistance. Clin Microbiol Rev 2012; 25(4): 661-81.
[http://dx.doi.org/10.1128/CMR.00043-12] [PMID: 23034325]

[21]   Ruppé É, Woerther PL, Barbier F. Mechanisms of antimicrobial resistance in Gram-negative bacilli. Ann Intensive Care 2015; 5(1): 61.
[http://dx.doi.org/10.1186/s13613-015-0061-0] [PMID: 26261001]

[22]   Paterson DL. Resistance in gram-negative bacteria: *enterobacteriaceae.* Am J Med 2006; 119(6) (Suppl. 1): S20-8.
[http://dx.doi.org/10.1016/j.amjmed.2006.03.013] [PMID: 16735147]

[23]   Peirano G, Pitout JDD. Extended-Spectrum β-Lactamase-Producing *Enterobacteriaceae*: Update on Molecular Epidemiology and Treatment Options. Drugs 2019; 79(14): 1529-41.
[http://dx.doi.org/10.1007/s40265-019-01180-3] [PMID: 31407238]

[24]   Ambler RP. The structure of beta-lactamases. Philos Trans R Soc Lond B Biol Sci 1980; 289(1036): 321-31.

[http://dx.doi.org/10.1098/rstb.1980.0049] [PMID: 6109327]

[25]   Bush K, Jacoby GA, Medeiros AA. A functional classification scheme for beta-lactamases and its correlation with molecular structure. Antimicrob Agents Chemother 1995; 39(6): 1211-33.
[http://dx.doi.org/10.1128/AAC.39.6.1211] [PMID: 7574506]

[26]   Pitout JDD, Laupland KB. Extended-spectrum beta-lactamase-producing *Enterobacteriaceae*: an emerging public-health concern. Lancet Infect Dis 2008; 8(3): 159-66.
[http://dx.doi.org/10.1016/S1473-3099(08)70041-0] [PMID: 18291338]

[27]   de Champs C, Sirot D, Chanal C, Poupart MC, Dumas MP, Sirot J. Concomitant dissemination of three extended-spectrum beta-lactamases among different *Enterobacteriaceae* isolated in a French hospital. J Antimicrob Chemother 1991; 27(4): 441-57.
[http://dx.doi.org/10.1093/jac/27.4.441] [PMID: 1856124]

[28]   Tzouvelekis LS, Tzelepi E, Tassios PT, Legakis NJ. CTX-M-type beta-lactamases: an emerging group of extended-spectrum enzymes. Int J Antimicrob Agents 2000; 14(2): 137-42.
[http://dx.doi.org/10.1016/S0924-8579(99)00165-X] [PMID: 10720804]

[29]   Lahlaoui H, Ben Haj Khalifa A, Ben Moussa M. Epidemiology of *Enterobacteriaceae* producing CTX-M type extended spectrum β-lactamase (ESBL). Med Mal Infect 2014; 44(9): 400-4.
[http://dx.doi.org/10.1016/j.medmal.2014.03.010] [PMID: 25234380]

[30]   Osano E, Arakawa Y, Wacharotayankun R, *et al.* Molecular characterization of an enterobacterial metallo beta-lactamase found in a clinical isolate of *Serratia marcescens* that shows imipenem resistance. Antimicrob Agents Chemother 1994; 38(1): 71-8.
[http://dx.doi.org/10.1128/AAC.38.1.71] [PMID: 8141584]

[31]   Lauretti L, Riccio ML, Mazzariol A, *et al.* Cloning and characterization of blaVIM, a new integron-borne metallo-beta-lactamase gene from a *Pseudomonas aeruginosa* clinical isolate. Antimicrob Agents Chemother 1999; 43(7): 1584-90.
[http://dx.doi.org/10.1128/AAC.43.7.1584] [PMID: 10390207]

[32]   Yigit H, Queenan AM, Anderson GJ, *et al.* Novel carbapenem-hydrolyzing beta-lactamase, KPC-1, from a carbapenem-resistant strain of *Klebsiella pneumoniae*. Antimicrob Agents Chemother 2001; 45(4): 1151-61.
[http://dx.doi.org/10.1128/AAC.45.4.1151-1161.2001] [PMID: 11257029]

[33]   Potron A, Poirel L, Rondinaud E, Nordmann P. Intercontinental spread of OXA-48 beta-lactamas-producing *Enterobacteriaceae* over a 11-year period, 2001 to 2011. Euro Surveill 2013; 18(31): 20549.
[http://dx.doi.org/10.2807/1560-7917.ES2013.18.31.20549] [PMID: 23929228]

[34]   Yong D, Toleman MA, Giske CG, *et al.* Characterization of a new metallo-beta-lactamase gene, bla(NDM-1), and a novel erythromycin esterase gene carried on a unique genetic structure in Klebsiella pneumoniae sequence type 14 from India. Antimicrob Agents Chemother 2009; 53(12): 5046-54.
[http://dx.doi.org/10.1128/AAC.00774-09] [PMID: 19770275]

[35]   Nordmann P, Poirel L, Walsh TR, Livermore DM. The emerging NDM carbapenemases. Trends Microbiol 2011; 19(12): 588-95.
[http://dx.doi.org/10.1016/j.tim.2011.09.005] [PMID: 22078325]

[36]   Boucher HW, Talbot GH, Bradley JS, *et al.* Bad bugs, no drugs: no ESKAPE! An update from the Infectious Diseases Society of America. Clin Infect Dis 2009; 48(1): 1-12.
[http://dx.doi.org/10.1086/595011] [PMID: 19035777]

[37]   Towner KJ. Acinetobacter: an old friend, but a new enemy. J Hosp Infect 2009; 73(4): 355-63.
[http://dx.doi.org/10.1016/j.jhin.2009.03.032] [PMID: 19700220]

[38]   Cai Y, Chai D, Wang R, Liang B, Bai N. Colistin resistance of *Acinetobacter baumannii*: clinical reports, mechanisms and antimicrobial strategies. J Antimicrob Chemother 2012; 67(7): 1607-15.

[http://dx.doi.org/10.1093/jac/dks084] [PMID: 22441575]

[39]   Woodford N, Turton JF, Livermore DM. Multiresistant Gram-negative bacteria: the role of high-risk clones in the dissemination of antibiotic resistance. FEMS Microbiol Rev 2011; 35(5): 736-55.
       [http://dx.doi.org/10.1111/j.1574-6976.2011.00268.x] [PMID: 21303394]

[40]   Wachino J, Arakawa Y. Exogenously acquired 16S rRNA methyltransferases found in aminoglycoside-resistant pathogenic Gram-negative bacteria: an update. Drug Resist Updat 2012; 15(3): 133-48.
       [http://dx.doi.org/10.1016/j.drup.2012.05.001] [PMID: 22673098]

[41]   Higgins PG, Pérez-Llarena FJ, Zander E, Fernández A, Bou G, Seifert H. OXA-235, a novel class D β-lactamase involved in resistance to carbapenems in *Acinetobacter baumannii*. Antimicrob Agents Chemother 2013; 57(5): 2121-6.
       [http://dx.doi.org/10.1128/AAC.02413-12] [PMID: 23439638]

[42]   Mugnier PD, Poirel L, Naas T, Nordmann P. Worldwide dissemination of the blaOXA-23 carbapenemase gene of *Acinetobacter baumannii*. Emerg Infect Dis 2010; 16(1): 35-40.
       [http://dx.doi.org/10.3201/eid1601.090852] [PMID: 20031040]

[43]   Adams-Haduch JM, Paterson DL, Sidjabat HE, *et al.* Genetic basis of multidrug resistance in *Acinetobacter baumannii* clinical isolates at a tertiary medical center in Pennsylvania. Antimicrob Agents Chemother 2008; 52(11): 3837-43.
       [http://dx.doi.org/10.1128/AAC.00570-08] [PMID: 18725452]

[44]   Hujer KM, Hamza NS, Hujer AM, *et al.* Identification of a new allelic variant of the Acinetobacter baumannii cephalosporinase, ADC-7 beta-lactamase: defining a unique family of class C enzymes. Antimicrob Agents Chemother 2005; 49(7): 2941-8.
       [http://dx.doi.org/10.1128/AAC.49.7.2941-2948.2005] [PMID: 15980372]

[45]   Tian GB, Adams-Haduch JM, Taracila M, Bonomo RA, Wang HN, Doi Y. Extended-spectrum AmpC cephalosporinase in Acinetobacter baumannii: ADC-56 confers resistance to cefepime. Antimicrob Agents Chemother 2011; 55(10): 4922-5.
       [http://dx.doi.org/10.1128/AAC.00704-11] [PMID: 21788456]

[46]   Fernández-Cuenca F, Martínez-Martínez L, Conejo MC, Ayala JA, Perea EJ, Pascual A. Relationship between beta-lactamase production, outer membrane protein and penicillin-binding protein profiles on the activity of carbapenems against clinical isolates of *Acinetobacter baumannii*. J Antimicrob Chemother 2003; 51(3): 565-74.
       [http://dx.doi.org/10.1093/jac/dkg097] [PMID: 12615856]

[47]   Krizova L, Poirel L, Nordmann P, Nemec A. TEM-1 β-lactamase as a source of resistance to sulbactam in clinical strains of *Acinetobacter baumannii*. J Antimicrob Chemother 2013; 68(12): 2786-91.
       [http://dx.doi.org/10.1093/jac/dkt275] [PMID: 23838947]

[48]   Houang ET, Chu YW, Lo WS, Chu KY, Cheng AF. Epidemiology of rifampin ADP-ribosyltransferase (arr-2) and metallo-beta-lactamase (blaIMP-4) gene cassettes in class 1 integrons in Acinetobacter strains isolated from blood cultures in 1997 to 2000. Antimicrob Agents Chemother 2003; 47(4): 1382-90.
       [http://dx.doi.org/10.1128/AAC.47.4.1382-1390.2003] [PMID: 12654674]

[49]   Coyne S, Courvalin P, Périchon B. Efflux-mediated antibiotic resistance in Acinetobacter spp. Antimicrob Agents Chemother 2011; 55(3): 947-53.
       [http://dx.doi.org/10.1128/AAC.01388-10] [PMID: 21173183]

[50]   Beceiro A, Llobet E, Aranda J, *et al.* Phosphoethanolamine modification of lipid A in colistin-resistant variants of *Acinetobacter baumannii* mediated by the pmrAB two-component regulatory system. Antimicrob Agents Chemother 2011; 55(7): 3370-9.
       [http://dx.doi.org/10.1128/AAC.00079-11] [PMID: 21576434]

[51]   Kim Y, Bae IK, Jeong SH, Yong D, Lee K. *In vivo* selection of pan-drug resistant acinetobacter

baumannii during antibiotic treatment. Yonsei Med J 2015; 56(4): 928-34.
[http://dx.doi.org/10.3349/ymj.2015.56.4.928] [PMID: 26069113]

[52]   Kaufmann SHE, Dorhoi A, Hotchkiss RS, Bartenschlager R. Host-directed therapies for bacterial and viral infections. Nat Rev Drug Discov 2018; 17(1): 35-56.
[http://dx.doi.org/10.1038/nrd.2017.162] [PMID: 28935918]

[53]   Zumla A, Rao M, Wallis RS, *et al.* Host-Directed Therapies Network consortium. Host-directed therapies for infectious diseases: current status, recent progress, and future prospects. Lancet Infect Dis 2016; 16(4): e47-63.
[http://dx.doi.org/10.1016/S1473-3099(16)00078-5] [PMID: 27036359]

[54]   Wallis RS, Maeurer M, Mwaba P, *et al.* Tuberculosis--advances in development of new drugs, treatment regimens, host-directed therapies, and biomarkers. Lancet Infect Dis 2016; 16(4): e34-46.
[http://dx.doi.org/10.1016/S1473-3099(16)00070-0] [PMID: 27036358]

[55]   Torfs E, Piller T, Cos P, Cappoen D. Opportunities for overcoming *Mycobacterium tuberculosis* drug resistance: emerging mycobacterial targets and host-directed therapy. Int J Mol Sci 2019; 20(12): 2868.
[http://dx.doi.org/10.3390/ijms20122868] [PMID: 31212777]

[56]   Andersson JA, Sha J, Kirtley ML, *et al.* Combating multidrug-resistant pathogens with host-directed nonantibiotic therapeutics. Antimicrob Agents Chemother 2017; 62(1): e01943-17.
[http://dx.doi.org/10.1128/AAC.01943-17] [PMID: 29109161]

[57]   Jatana S, Homer CR, Madajka M, *et al.* Pyrimidine synthesis inhibition enhances cutaneous defenses against antibiotic resistant bacteria through activation of NOD2 signaling. Sci Rep 2018; 8(1): 8708.
[http://dx.doi.org/10.1038/s41598-018-27012-0] [PMID: 29880914]

[58]   Guo X, Wang C, Xu T, Yang L, Liu C, Qi X. SiO_2 prompts host defense against Acinetobacter baumannii infection by mTORC1 activation. Sci China Life Sci 2020; 63
[http://dx.doi.org/10.1007/s11427-020-1781-8] [PMID: 32880864]

[59]   Abdulnour RE, Sham HP, Douda DN, *et al.* Aspirin-triggered resolvin D1 is produced during self-resolving gram-negative bacterial pneumonia and regulates host immune responses for the resolution of lung inflammation. Mucosal Immunol 2016; 9(5): 1278-87.
[http://dx.doi.org/10.1038/mi.2015.129] [PMID: 26647716]

[60]   Chauhan A, Lebeaux D, Ghigo JM, Beloin C. Full and broad-spectrum *in vivo* eradication of catheter-associated biofilms using gentamicin-EDTA antibiotic lock therapy. Antimicrob Agents Chemother 2012; 56(12): 6310-8.
[http://dx.doi.org/10.1128/AAC.01606-12] [PMID: 23027191]

[61]   Yenn TW, Arslan Khan M, Amiera Syuhada N, Chean Ring L, Ibrahim D, Tan WN. Stigmasterol: An adjuvant for beta lactam antibiotics against beta-lactamase positive clinical isolates. Steroids 2017; 128: 68-71.
[http://dx.doi.org/10.1016/j.steroids.2017.10.016] [PMID: 29104098]

[62]   Ferrer A, Altabella T, Arró M, Boronat A. Emerging roles for conjugated sterols in plants. Prog Lipid Res 2017; 67: 27-37.
[http://dx.doi.org/10.1016/j.plipres.2017.06.002] [PMID: 28666916]

[63]   Zurawski DV, Reinhart AA, Alamneh YA, *et al.* SPR741, an antibiotic adjuvant, potentiates the *in vitro* and *in vivo* activity of rifampin against clinically relevant extensively drug-resistant *Acinetobacter baumannii*. Antimicrob Agents Chemother 2017; 61(12): e01239-17.
[http://dx.doi.org/10.1128/AAC.01239-17] [PMID: 28947471]

[64]   Corbett D, Wise A, Langley T, *et al.* Potentiation of antibiotic activity by a novel cationic peptide: potency and spectrum of activity of SPR741. Antimicrob Agents Chemother 2017; 61(8): e00200-17.
[http://dx.doi.org/10.1128/AAC.00200-17] [PMID: 28533232]

[65]   Durand-Réville TF, Guler S, Comita-Prevoir J, *et al.* ETX2514 is a broad-spectrum β-lactamase

inhibitor for the treatment of drug-resistant Gram-negative bacteria including Acinetobacter baumannii. Nat Microbiol 2017; 2: 17104.
[http://dx.doi.org/10.1038/nmicrobiol.2017.104] [PMID: 28665414]

[66]   Baker KR, Jana B, Hansen AM, *et al.* Repurposing azithromycin and rifampicin against Gram-negative pathogens by combination with peptide potentiators. Int J Antimicrob Agents 2019; 53(6): 868-72.
[http://dx.doi.org/10.1016/j.ijantimicag.2018.10.025] [PMID: 30447380]

[67]   Song M, Liu Y, Huang X, *et al.* A broad-spectrum antibiotic adjuvant reverses multidrug-resistant Gram-negative pathogens. Nat Microbiol 2020; 5(8): 1040-50.
[http://dx.doi.org/10.1038/s41564-020-0723-z] [PMID: 32424338]

[68]   Douafer H, Andrieu V, Wafo E, Brunel JM. Characterization of a new aerosol antibiotic/adjuvant combination for the treatment of *P. aeruginosa* lung infections. Int J Pharm 2020; 586: 119548.
[http://dx.doi.org/10.1016/j.ijpharm.2020.119548] [PMID: 32565286]

[69]   Barker WT, Jania LA, Melander RJ, Koller BH, Melander C. Eukaryotic phosphatase inhibitors enhance colistin efficacy in gram-negative bacteria. Chem Biol Drug Des 2020; 96(5): 1180-6.
[http://dx.doi.org/10.1111/cbdd.13735] [PMID: 32562384]

[70]   Schmid A, Wolfensberger A, Nemeth J, Schreiber PW, Sax H, Kuster SP. Monotherapy *versus* combination therapy for multidrug-resistant Gram-negative infections: systematic review and meta-analysis. Sci Rep 2019; 9(1): 15290.
[http://dx.doi.org/10.1038/s41598-019-51711-x] [PMID: 31664064]

[71]   Munoz-Price LS, Poirel L, Bonomo RA, *et al.* Clinical epidemiology of the global expansion of *Klebsiella pneumoniae* carbapenemases. Lancet Infect Dis 2013; 13(9): 785-96.
[http://dx.doi.org/10.1016/S1473-3099(13)70190-7] [PMID: 23969216]

[72]   Tzouvelekis LS, Markogiannakis A, Piperaki E, Souli M, Daikos GL. Treating infections caused by carbapenemase-producing *Enterobacteriaceae.* Clin Microbiol Infect 2014; 20(9): 862-72.
[http://dx.doi.org/10.1111/1469-0691.12697] [PMID: 24890393]

[73]   Pitout JD, Nordmann P, Poirel L. Carbapenemase-producing *Klebsiella pneumoniae*, a key pathogen set for global nosocomial dominance. Antimicrob Agents Chemother 2015; 59(10): 5873-84.
[http://dx.doi.org/10.1128/AAC.01019-15] [PMID: 26169401]

[74]   Candan ED, Aksöz N. *Klebsiella pneumoniae:* characteristics of carbapenem resistance and virulence factors. Acta Biochim Pol 2015; 62(4): 867-74.
[http://dx.doi.org/10.18388/abp.2015_1148] [PMID: 26637376]

[75]   Daikos GL, Tsaousi S, Tzouvelekis LS, *et al.* Carbapenemase-producing *Klebsiella pneumoniae* bloodstream infections: lowering mortality by antibiotic combination schemes and the role of carbapenems. Antimicrob Agents Chemother 2014; 58(4): 2322-8.
[http://dx.doi.org/10.1128/AAC.02166-13] [PMID: 24514083]

[76]   Tumbarello M, Trecarichi EM, De Rosa FG, *et al.* ISGRI-SITA (Italian Study Group on Resistant Infections of the Società Italiana Terapia Antinfettiva). Infections caused by KPC-producing *Klebsiella pneumoniae*: differences in therapy and mortality in a multicentre study. J Antimicrob Chemother 2015; 70(7): 2133-43.
[http://dx.doi.org/10.1093/jac/dkv086] [PMID: 25900159]

[77]   Gomez-Simmonds A, Nelson B, Eiras DP, *et al.* Combination regimens for treatment of carbapenem-resistant Klebsiella pneumoniae bloodstream infections. Antimicrob Agents Chemother 2016; 60(6): 3601-7.
[http://dx.doi.org/10.1128/AAC.03007-15] [PMID: 27044555]

[78]   Abdelsalam MFA, Abdalla MS, El-Abhar HSE. Prospective, comparative clinical study between high-dose colistin monotherapy and colistin-meropenem combination therapy for treatment of hospital-acquired pneumonia and ventilator-associated pneumonia caused by multidrug-resistant *Klebsiella pneumoniae.* J Glob Antimicrob Resist 2018; 15: 127-35.

[http://dx.doi.org/10.1016/j.jgar.2018.07.003] [PMID: 30010061]

[79] Paul M, Daikos GL, Durante-Mangoni E, *et al.* Colistin alone *versus* colistin plus meropenem for treatment of severe infections caused by carbapenem-resistant Gram-negative bacteria: an open-label, randomised controlled trial. Lancet Infect Dis 2018; 18(4): 391-400.
[http://dx.doi.org/10.1016/S1473-3099(18)30099-9] [PMID: 29456043]

[80] Wunderink RG, Giamarellos-Bourboulis EJ, Rahav G, *et al.* effect and safety of meropenem-vaborbactam *versus* best-available therapy in patients with carbapenem-resistant *Enterobacteriaceae* infections: The TANGO II randomized clinical trial. Infect Dis Ther 2018; 7(4): 439-55.
[http://dx.doi.org/10.1007/s40121-018-0214-1] [PMID: 30270406]

[81] Ji S, Lv F, Du X, *et al.* Cefepime combined with amoxicillin/clavulanic acid: a new choice for the KPC-producing *K. pneumoniae* infection. Int J Infect Dis 2015; 38: 108-14.
[http://dx.doi.org/10.1016/j.ijid.2015.07.024] [PMID: 26255892]

[82] De Pascale G, Martucci G, Montini L, *et al.* Double carbapenem as a rescue strategy for the treatment of severe carbapenemase-producing *Klebsiella pneumoniae* infections: a two-center, matched case-control study. Crit Care 2017; 21(1): 173.
[http://dx.doi.org/10.1186/s13054-017-1769-z] [PMID: 28679413]

[83] Bandick RG, Mousavi S, Bereswill S, Heimesaat MM. Review of therapeutic options for infections with carbapenem-resistant *Klebsiella pneumoniae*. Eur J Microbiol Immunol (Bp) 2020; 10: 115-24.
[http://dx.doi.org/10.1556/1886.2020.00022] [PMID: 32946421]

[84] Pang Z, Raudonis R, Glick BR, Lin TJ, Cheng Z. Antibiotic resistance in *Pseudomonas aeruginosa*: mechanisms and alternative therapeutic strategies. Biotechnol Adv 2019; 37(1): 177-92.
[http://dx.doi.org/10.1016/j.biotechadv.2018.11.013] [PMID: 30500353]

[85] Solomkin J, Hershberger E, Miller B, *et al.* Ceftolozane/Tazobactam plus metronidazole for complicated intra-abdominal infections in an era of multidrug resistance: results from a randomized, double-blind, phase 3 trial (ASPECT-cIAI). Clin Infect Dis 2015; 60(10): 1462-71.
[http://dx.doi.org/10.1093/cid/civ097] [PMID: 25670823]

[86] Huntington JA, Sakoulas G, Umeh O, *et al.* Efficacy of ceftolozane/tazobactam *versus* levofloxacin in the treatment of complicated urinary tract infections (cUTIs) caused by levofloxacin-resistant pathogens: results from the ASPECT-cUTI trial. J Antimicrob Chemother 2016; 71(7): 2014-21.
[http://dx.doi.org/10.1093/jac/dkw053] [PMID: 26994090]

[87] Kollef MH, Nováček M, Kivistik Ü, *et al.* Ceftolozane-tazobactam *versus* meropenem for treatment of nosocomial pneumonia (ASPECT-NP): a randomised, controlled, double-blind, phase 3, non-inferiority trial. Lancet Infect Dis 2019; 19(12): 1299-311.
[http://dx.doi.org/10.1016/S1473-3099(19)30403-7] [PMID: 31563344]

[88] Bassetti M, Castaldo N, Cattelan A, *et al.* CEFTABUSE Study Group. Ceftolozane/tazobactam for the treatment of serious *Pseudomonas aeruginosa* infections: a multicentre nationwide clinical experience. Int J Antimicrob Agents 2019; 53(4): 408-15.
[http://dx.doi.org/10.1016/j.ijantimicag.2018.11.001] [PMID: 30415002]

[89] Stone GG, Newell P, Gasink LB, *et al.* Clinical activity of ceftazidime/avibactam against MDR *Enterobacteriaceae* and *Pseudomonas aeruginosa*: pooled data from the ceftazidime/avibactam Phase III clinical trial programme. J Antimicrob Chemother 2018; 73(9): 2519-23.
[http://dx.doi.org/10.1093/jac/dky204] [PMID: 29912399]

[90] Torres A, Zhong N, Pachl J, *et al.* Ceftazidime-avibactam *versus* meropenem in nosocomial pneumonia, including ventilator-associated pneumonia (REPROVE): a randomised, double-blind, phase 3 non-inferiority trial. Lancet Infect Dis 2018; 18(3): 285-95.
[http://dx.doi.org/10.1016/S1473-3099(17)30747-8] [PMID: 29254862]

[91] Pogue JM, Kaye KS, Veve MP, *et al.* Ceftolozane/Tazobactam vs polymyxin or aminoglycoside-based regimens for the treatment of drug-resistant *Pseudomonas aeruginosa*. Clin Infect Dis 2020; 71(2): 304-10.

[http://dx.doi.org/10.1093/cid/ciz816] [PMID: 31545346]

[92]   Lupo A, Haenni M, Madec JY. Antimicrobial resistance in *Acinetobacter* spp. and *Pseudomonas* spp. Microbiol Spectr 2018; 6(3)
       [http://dx.doi.org/10.1128/microbiolspec.ARBA-0007-2017] [PMID: 30101740]

[93]   Aydemir H, Akduman D, Piskin N, *et al.* Colistin vs. the combination of colistin and rifampicin for the treatment of carbapenem-resistant *Acinetobacter baumannii* ventilator-associated pneumonia. Epidemiol Infect 2013; 141(6): 1214-22.
       [http://dx.doi.org/10.1017/S095026881200194X] [PMID: 22954403]

[94]   Durante-Mangoni E, Signoriello G, Andini R, *et al.* Colistin and rifampicin compared with colistin alone for the treatment of serious infections due to extensively drug-resistant *Acinetobacter baumannii*: a multicenter, randomized clinical trial. Clin Infect Dis 2013; 57(3): 349-58.
       [http://dx.doi.org/10.1093/cid/cit253] [PMID: 23616495]

[95]   Sirijatuphat R, Thamlikitkul V. Preliminary study of colistin *versus* colistin plus fosfomycin for treatment of carbapenem-resistant *Acinetobacter baumannii* infections. Antimicrob Agents Chemother 2014; 58(9): 5598-601.
       [http://dx.doi.org/10.1128/AAC.02435-13] [PMID: 24982065]

[96]   Qin Y, Zhang J, Wu L, Zhang D, Fu L, Xue X. Comparison of the treatment efficacy between tigecycline plus high-dose cefoperazone-sulbactam and tigecycline monotherapy against ventilator-associated pneumonia caused by extensively drug-resistant *Acinetobacter baumannii*. Int J Clin Pharmacol Ther 2018; 56(3): 120-9.
       [http://dx.doi.org/10.5414/CP203102] [PMID: 29319497]

[97]   Mosaed R, Haghighi M, Kouchak M, *et al.* Interim Study: Comparison of safety and efficacy of levofloxacin plus colistin regimen with levofloxacin plus high dose ampicillin/sulbactam infusion in treatment of ventilator-associated pneumonia due to multi drug resistant Acinetobacter. Iran J Pharm Res 2018; 17 (Suppl. 2): 206-13.
       [PMID: 31011353]

[98]   Khalili H, Shojaei L, Mohammadi M, Beigmohammadi MT, Abdollahi A, Doomanlou M. Meropenem/colistin *versus* meropenem/ampicillin-sulbactam in the treatment of carbapenem-resistant pneumonia. J Comp Eff Res 2018; 7(9): 901-11.
       [http://dx.doi.org/10.2217/cer-2018-0037] [PMID: 30192166]

[99]   Pourheidar E, Haghighi M, Kouchek M, *et al.* comparison of intravenous ampicillin-sulbactam plus nebulized colistin with intravenous colistin plus nebulized colistin in treatment of ventilator associated pneumonia caused by multi drug resistant *Acinetobacter baumannii*: Randomized open label trial. Iran J Pharm Res 2019; 18 (Suppl. 1): 269-81.
       [PMID: 32802106]

[100] Harris PNA, Tambyah PA, Lye DC, *et al.* MERINO Trial Investigators and the Australasian Society for Infectious Disease Clinical Research Network (ASID-CRN). effect of piperacillin-tazobactam vs meropenem on 30-day mortality for patients with *E. coli* or *Klebsiella pneumoniae* bloodstream infection and ceftriaxone resistance: a randomized clinical trial. JAMA 2018; 320(10): 984-94.
       [http://dx.doi.org/10.1001/jama.2018.12163] [PMID: 30208454]

[101] Sims M, Mariyanovski V, McLeroth P, *et al.* Prospective, randomized, double-blind, Phase 2 dose-ranging study comparing efficacy and safety of imipenem/cilastatin plus relebactam with imipenem/cilastatin alone in patients with complicated urinary tract infections. J Antimicrob Chemother 2017; 72(9): 2616-26.
       [http://dx.doi.org/10.1093/jac/dkx139] [PMID: 28575389]

[102] Kaye KS, Bhowmick T, Metallidis S, *et al.* Effect of Meropenem-Vaborbactam *vs* Piperacillin-Tazobactam on clinical cure or improvement and microbial eradication in complicated urinary tract infection: The TANGO I Randomized Clinical Trial. JAMA 2018; 319(8): 788-99.
[http://dx.doi.org/10.1001/jama.2018.0438] [PMID: 29486041]

CHAPTER 3

Bioactive Substances as Anti-infective Strategies Against *Clostridioides Difficile*

Joana Barbosa[1,*] and **Paula Teixeira**[1]

[1] *Universidade Católica Portuguesa, CBQF - Centro de Biotecnologia e Química Fina—Laboratório Associado, Escola Superior de Biotecnologia, Porto, Portugal*

Abstract: The incidence and severity of diarrhea associated with *Clostridioides difficile* increased exponentially worldwide until 2004. But during the last few years, a downward trend has been observed globally, except in some European countries and Asia.

Until recently, *C. difficile* was the primary cause of nosocomial infection following antibiotic exposure, presenting a high rate of mortality and morbidity. However, the emergence and spread of a hypervirulent strain (BI/NAP1/027) and an increase in the incidence of community-acquired *C. difficile* infection (CDI), especially in populations not previously considered at high risk, have contributed to alterations in the infection epidemiology. After initial treatment with broad-spectrum antibiotics, CDI recurrence is the cause of substantial morbidity, indicating that alternative strategies to the usual therapeutics are urgently needed.

Several studies have investigated probiotics to assess their preventative and/or prophylactic effects on CDI, but their use is still controversial. Other anti-infective alternatives, such as bacteriocins and phage therapy, appear as promising answers for CDI treatment.

This review explores the current therapy approaches and the advances in searching for alternative solutions to inhibit the opportunistic pathogen *C. difficile*.

Keywords: Antibiotics, Antimicrobials, Bacteriocins, Bacteriophages, Clostridioides difficile infection (CDI), Fecal microbiota transplantation, Fidaxomicin, Monoclonal antibodies, Metronidazole, Non-toxigenic spores, Probiotics, Recurrent CDI, Synthetic polymers, Vaccines, Vancomycin.

---

* **Corresponding author Joana Barbosa:** Universidade Católica Portuguesa, CBQF - Centro de Biotecnologia e Química Fina—Laboratório Associado, Escola Superior de Biotecnologia; Rua Diogo Botelho 1327, 4169-005 Porto, Portugal, Tel: +351225580001; Fax: +351225090351; E-mail: jbarbosa@porto.ucp.pt

INTRODUCTION

Clostridioides difficile (previously *Clostridium difficile*) [1] is a Gram-positive, endospore-forming, and toxin-producing anaerobic species. It was originally isolated from neonates' stools in 1935 [2] as an inoffensive inhabitant of commensal intestinal microbiota. However, in 1978 *C. difficile* was recognized as an important cause of antimicrobial-associated diarrhea in hospitalized individuals [3].

Spores of *C. difficile* can be found ubiquitously in the environment. Once ingested, these highly resistant spores can survive to the gastrointestinal tract barriers and remain inactive, resisting the host's immune system mechanisms. *Clostridioides difficile* produces major toxins responsible for mild-to-severe forms of gastrointestinal infections, ranging from asymptomatic intestinal colonization, self-limiting mild diarrhea to severe or life-threatening pseudomembranous colitis, toxic megacolon, sepsis, and death [4, 5]. The main predisposing factor for developing *C. difficile* infection (CDI) is antibiotic therapy, particularly with broad-spectrum antibiotics [6]. The normal gastrointestinal microbiome acts as a colonization barrier, preventing the spore's germination and their return to a toxin-producing vegetative state [5]. Disturbance of the gastrointestinal microbiome due to antibiotic exposure allows spore germination and vegetative growth of *C. difficile* [6]. Vegetative cells enter the mucus layer and adhere to intestinal epithelial cells, proliferating and colonizing the large intestine, where they produce and release toxins [7].

Gut-targeted therapies for CDI have a relatively direct effect on a mild disease. Nevertheless, the same is not often verified in patients with severe disease [8], and recurrence of CDI occurs in many patients after initial treatment with broad-spectrum antibiotics [9]. Since the main clinical challenges are the recurrence of CDI and the resistance to antibiotics currently used in the therapy, alternative therapeutic strategies are urgently needed.

This chapter intends to review the therapy approaches recommended for CDI treatment and explore the advances in searching for new therapies, emphasizing natural alternatives to currently used antibiotics to inhibit the opportunistic pathogen *C. difficile*.

PATHOGENESIS AND EPIDEMIOLOGY OF *C. DIFFICILE*

The main virulence factor mediating the pathogenesis of *C. difficile* disease is the production of two large toxins, toxin A and toxin B [10, 11], encoded by *tcdA* and *tcdB* genes, respectively. These are located within a region of the chromosome identified as pathogenicity locus or PaLoc [11]. Some strains can also produce a binary toxin, *C. difficile* transferase (CDT) [12]. This toxin is encoded by *cdtA* and *cdtB* genes that are located outside the PaLoc region at the part of the binary toxin locus in the genome (CdtLoc) [13]. Although there is no evidence that it causes disease *per se*, this binary toxin has been associated with more severe disease [14].

Epidemiology of *C. difficile* has been changing since the mid-2000s; community-associated infections have become more frequent among younger and relatively healthy individuals due to unknown predisposing factors, antibiotic therapy, and previous hospitalization status [15, 16]. The epidemiology of *C. difficile* in children is characterized by the transient colonization of different toxigenic and non-toxigenic strains at different times [17] and, despite the severe forms of CDI are still less common in children [18, 19], rates of recurrent CDI are similar to those in adults.

The emergence and spread of a new hypervirulent *C. difficile* BI/NAP1/027 strain (characterized as BI group by restriction endonuclease analysis, North American pulsed-field type NAP1 by pulse-field gel electrophoresis, and ribotype 027 by polymerase chain reaction ribotyping) [20] may be correlated with the new epidemiology of CDI, due to its increased sporulation [21] or hyperexpression of toxins [22]. The mechanisms by which transmission occurs in the community are not yet known.

CURRENT APPROACHES IN THE TREATMENT OF CDI

Antibiotic therapy is the treatment of choice for CDI. The severity of the disease dictates the specific treatment to be used, based on guideline recommendations [17, 23]. Until recently, vancomycin (bacterial cell wall synthesis inhibition) and metronidazole (inhibition of deoxyribonucleic acid synthesis and DNA degradation) were the first-choice antibiotics. However, recurrent CDI with significant morbidity and mortality upon the cessation of their administration triggered the development of a new and specific RNA synthesis inhibitor antibiotic, fidaxomicin (previously designated as OPT-80) [24]. The use of fidaxomicin was approved in adults (May 2011) and in children (January 2020) by the US Food and Drug Administration (FDA). It does not cause significant changes to the intestinal microbiome of infected patients during [25] and after treatment [26] compared to vancomycin, resulting in lower rates of relapse. Additionally, treatment with fidaxomicin showed reduced acquisition and overgrowth of vancomycin-resistant enterococci and *Candida* spp [27], unlike the treatment with vancomycin. Frequent use of vancomycin should be avoided since it leads to increased colonization of vancomycin-resistant enterococci, increasing the CDI recurrence risk [28]. In addition to the lower cure rates in severe CDI than vancomycin and fidaxomicin, metronidazole should also be avoided for repeated and long periods due to its several adverse effects, including potential cumulative neurotoxicity [29].

For the prevention/reduction of the CDI risk, the frequency and duration of high-risk antibiotic therapy for CDI should be minimized, and an antibiotic administration program should be implemented [17, 23]. A summary of the treatment guideline for both adults and children is shown in Fig. (1).

	Initial episode of non-severe CDI	Initial episode of severe CDI	Initial episode of fulminant CDI	First recurrence of CDI	Second and subsequent recurrences
ADULTS	VAN (125 mg orally 4x/day) 10 days **OR** FDX (200 mg 2x/day) 10 days **OR** MTZ (500 mg orally 3x/day) 10 days, only if VAN and FXM were inaccessible	VAN (125 mg orally 4x/day) 10 days **OR** FDX (200 mg 2x/day) 10 days	VAN (500 mg orally 4x/day) **OR** VAN[1] (500 mg per rectum in about 100 ml normal saline every 6h) **OR** MTZ[1] (500 mg intravenously every 8h) plus VAN (oral or rectal)	FDX[2] (200 mg 2x/day) 10 days **OR** VAN[3] (tapered and pulsed regimen) **OR** VAN[4] (125 mg 4x/day)	VAN (tapered and pulsed regimen) **OR** VAN (125 mg orally 4x/day) 10 days followed rifaximin (400 mg 3x/day) 20 days **OR** FDX (200 mg 2x/day) 10 days **OR** FMT
CHILDREN	MTZ (7.5 mg/kg/dose orally 3 or 4x/day) 10 days **OR** VAN (10 mg/kg/dose orally 4x/day) 10 days	VAN (10 mg/kg/dose 4x/day, oral or rectal) 10 days with or without MTZ (10 mg/kg/dose intravenously 3x/day) 10 days	VAN (10 mg/kg/dose 4x/day, oral or rectal) 10 days with or without MTZ (10 mg/kg/dose intravenously 3x/day) 10 days	VAN* (10 mg/kg/dose orally 4x/day) 10 days **OR** MTZ (7.5 mg/kg/dose orally 3 or 4x/day) 10 days	VAN (orally 10 mg/kg/dose; max. dose of 125 mg) in tapered and pulse regimen for 10 days **OR** VAN (orally 10 mg/kg/dose 4x/day) 10 days followed by rifaximin for 20 days[5] **OR** FMT

Fig. (1). Schematics of the treatment guidelines applied to *Clostridioides difficile* infection (CDI).

VAN: vancomycin, FDX: fidaxomicin, MTZ: metronidazole, FMT: fecal microbiota transplantation, [1]If ileus is present; [2]If VAN was used in an initial CDI episode; [3]If a standard regimen was used in an initial CDI episode; [4]If MTZ was used in an initial CDI episode; [5]Maximum dose of 400 mg three x/day and only for children >12 years old

Initial Episodes of CDI

In adults, the prescription for ten days of vancomycin (125 mg orally four times/day) or fidaxomicin (200 mg two times/day) is strongly recommended for an initial episode of both non-severe (white blood cell count ≤ 15000 cells/mL and a serum creatinine level < 1.5 mg/dL) and severe (white blood cell count ≥

15000 cells/mL or a serum creatinine level > 1.5 mg/dL) CDI. The use of metronidazole for ten days (500 mg orally three times/day) is suggested when vancomycin or fidaxomicin is inaccessible but merely for non-severe CDI episodes [17]. In a 21-day prospective, randomized, double-blinded, and placebo-controlled trial with 172 patients, Zar *et al.* found that for patients with severe CDI, treatment with vancomycin was superior to treatment with metronidazole with cure rates of 97% and 76%, respectively (p = 0.02) [8].

Metronidazole (7.5 mg/kg/dose orally three or four times/day) or vancomycin (10 mg/kg/dose orally four times/day) for ten days is also recommended for the treatment of children with an initial episode of non-severe CDI, but oral vancomycin (10 mg/kg/dose four times/day) is recommended over metronidazole for children with severe CDI [17]. With the FDA's recent approval, it is expected that the use of fidaxomicin will also be recommended. In a multi-center, investigator-blind, phase 3, parallel-group trial conducted by Wolf *et al.*, children and adolescents (< 18 years old) with confirmed CDI were randomized in a ratio of 2:1 to the treatment with fidaxomicin or vancomycin for ten days [30]. The authors reported that 142 out of 148 patients (30 with < 2 years old) were treated, and the global cure rate of participants receiving fidaxomicin (68.4%) was significantly higher compared to vancomycin (50.0%) [30].

Fulminant Episodes of CDI

Administration of vancomycin orally (500 mg four times/day), or per rectum (500 mg in about 100 mL normal saline every six hours) if ileus is present is recommended in adults for fulminant CDI, *i.e.*, for complicated CDI characterized by hypotension, shock, ileus or megacolon. Administration of metronidazole (500 mg intravenously every eight hours) together with oral or rectal vancomycin is also recommended if ileus is present. Subtotal colectomy with preservation of the rectum may be necessary for gravely ill patients [17]. For children, oral or rectal vancomycin (10 mg/kg/dose four times/day) is recommended for ten days, and intravenous metronidazole (10 mg/kg/dose three times/day) administered together with oral vancomycin should also be considered [17].

Recurrence of CDI

Recurrent CDI is usually defined as an episode of CDI occurring within two to eight weeks, following a previous episode after successfully completing treatments [31]. About 10% to 30% of patients with one CDI episode will experience recurrent CDI [32]. The risk of further recurrences significantly increases after the first recurrence [31, 33].

First Recurrence of CDI

If vancomycin was administered in an initial episode of CDI in adults, then the the use of fidaxomicin (200 mg orally two times/day) for ten days or vancomycin but only as a tapered and pulsed regimen (125 mg orally four times/day for 10-14 days, two times/day for one week, one time/day for one week and every two or three days for 2-8 weeks) is recommended. Administration of vancomycin in a standard regimen (125 mg four times/day) should be used only if an initial episode was treated with metronidazole [17]. For children, vancomycin (10 mg/kg/dose orally four times/day) for ten days is recommended over metronidazole (7.5 mg/kg/dose orally three or four times/day) [17].

Second or Subsequent Recurrence of CDI

Adult patients with more than one CDI recurrence should be treated with oral vancomycin in a tapered and pulsed regimen or for ten days with vancomycin (125 mg orally four times/day), followed by rifaximin (400 mg three times/day) or fidaxomicin (200 mg two times/day) for 20 and 10 days, respectively. For patients with several CDI recurrences and failure of the applied appropriate antibiotic therapy, fecal microbiota transplantation is strongly recommended [17].

Vancomycin (10 mg/kg/dose orally with a maximum dose of 125 mg) in a tapered and pulsed regimen is recommended for children with second or subsequent recurrence of CDI. Alternatively, vancomycin (10 mg/kg/dose four times/day) could be administered for ten days followed by rifaximin for 20 days (maximum dose of 400 mg three times/day and its use is not approved in children < 12 years old by FDA) or, despite the low quality of evidence, fecal microbiota transplantation could be considered [17].

Fecal Microbiota Transplantation

Fecal microbiota transplantation (FMT) is defined as an infusion of stool from a healthy individual transferred into a patient's lower gastrointestinal tract *via* nasogastric tube, colonoscopy or enema, to create a new gut microbiome community, restoring the normal gut function [34]. The effectiveness of FMT in recurrent CDI has been proven with successful clinical cure rates of 70 - 90%, with no related significant adverse events [33, 35].

Available reported cases suggested low-quality evidence of the efficacy of FMT in children [36, 37]. However, recently, Nicholson *et al.* conducted a multi-center retrospective cohort study of 372 patients (pediatric and young adults from 11 months to 23 years old) and concluded that FMT was safe and effective, *i.e.*, with no CDI recurrences after two months following FMT: 81.0% and 86.6% of the patients had a successful result following a single FMT and following a first or repeated FMT, respectively [38].

Although being effective and apparently safe, there are still many reservations about the long-term safety of fecal microbiota transplantation.

PROMISING CANDIDATE ANTIBIOTICS FOR THE PREVENTION/-TREATMENT OF CDI AS ALTERNATIVE TO STANDARD THERAPIES

The limited number of effective antibiotics used in the treatment of CDI and the increased failure in both clinical cure rates and recurrences, coupled with the increased resistance amongst the more virulent *C. difficile* strains, are of particular concern. Alternative therapies are scarce, but in recent years, new and safer drugs focusing on the prevention and treatment of CDI have been developed, although the majority need to be confirmed in phase III clinical trials.

Cadazolid

Cadazolid is a fluoroquinolone–oxazolidinone that inhibits protein and DNA synthesis, with recognized bactericidal activity against *C. difficile*, suppressing the synthesis of toxins A and B as well as spore formation in the main virulent strains [39]. In phase II clinical trials (NCT01222702 and NCT01222702), in addition to being safe and well-tolerated, cadazolid was also proved to exert higher clinical response and a lower recurrence rate per each dose of cadazolid compared to vancomycin [40, 41]. However, the development of this antibiotic was discontinued due to its clinical results in phase III randomized clinical trials (NCT01987895 and NCT01983683), *i.e.*, cadazolid was not superior to vancomycin for clinical cure [42]. Also, the only known prospective and multi-center trial to investigate the efficacy of cadazolid *versus* vancomycin in children was suspended in 2018 (NCT03105479). So, this promising candidate is no longer a viable strategy to control CDI.

CRS3123

CRS3123 (REP3123) is a fully synthetic diaryldiamine that inhibits protein synthesis preventing the growth and spore production of *C. difficile* [43]. In a phase I single-center, double-blind, placebo-controlled, and dose-escalation trial, 40 participants were randomized in cohorts of eight participants to receive different doses of CRS3123 (100, 200, 400, 800, and 1200 mg) or placebo. No adverse events were reported for doses up to 1200 mg CRS3123 demonstrating it to be a safe and well-tolerated antimicrobial [44].

The safety and efficacy of CRS3123 compared to vancomycin will be evaluated for the first time in patients with CDI in an ongoing phase II clinical trial (75N93019C00056). To date, no study with CRS3123 in children is being carried out.

Nitazoxanide

Nitazoxanide is a nitrothiazole benzamide initially FDA-approved for treating cryptosporidiosis and giardiasis in children ≥ one year of age and adults, which, at low concentrations, also inhibits *C. difficile in vitro* [45]. In phase III double-blind, prospective, and randomized study with 110 participants, despite nitazoxanide being safe and effective for initial treatment or recurrent CDI, similar clinical responses were observed for patients treated with nitazoxanide (500 mg for seven or ten days) or metronidazole (250 mg for ten days) [46]. Another phase III double-blind, prospective, and randomized study (NCT00384527) with 49 participants was conducted to compare nitazoxanide (500 mg two times/day) with vancomycin (125 mg four times/day) for ten days, but the study was ended early due to slow enrollment [47]. According to the Australasian Society of Infectious Diseases [48] guidelines, oral nitazoxanide is already recommended as an alternative treatment of recurrent CDI in children.

Ramoplanin

Ramoplanin is a lipoglycodepsipeptide that inhibits bacterial cell wall biosynthesis and has a bactericidal effect against *C. difficile in vitro* [49] and in animal models [50]. In a phase II study with 86 participants, similar clinical response rates were reported with ramoplanin 200 mg (77%), ramoplanin 400 mg (80%), and vancomycin 125 mg (80%) [51]. However, in another phase II study, two daily doses of ramoplanin (100 mg or 400 mg) were administered to 68 patients and, after 21 days of treatment, no differences were found between ramoplanin and placebo groups in vancomycin-resistant enterococci carriage, *i.e.*, ramoplanin was effective suppressing gastrointestinal carriage of vancomycin-resistant enterococci [52]. No other clinical trials are available for ramoplanin in adults or children.

Ridinilazole

Although little is known about the action of the antibiotic ridinilazole, it seems that it acts on cell division and has a bactericidal effect against *C. difficile* [53]. In a phase II clinical trial (NCT02092935), 100 participants aged between 18 and 90 years old received oral ridinilazole (200 mg two times/day) or oral vancomycin (125 mg four times/day) for ten days. A high potential for ridinilazole was shown, with a statistical superiority at the 10% level in the treatment of initial CDI and

CDI recurrence reduction compared to vancomycin [54]. Two phase III clinical trials are currently ongoing (NCT03595553 and NCT03595566) for patients aged 18 years and older.

Surotomycin

Surotomycin, a cyclic lipopeptide antibiotic formed by enzymatical cleavage of daptomycin, has shown a concentration-dependent bactericidal effect against *C. difficile* in both logarithmic- and stationary-phases [55]. Similar to the antibiotic cadazolid, surotomycin successfully completed a phase II clinical trial (NCT01085591) in adults with CDI demonstrating higher continued cure rates as well as the reduction and delay in CDI recurrence episodes in comparison with vancomycin [56] but failed in both phase III clinical trials conducted (NCT01597505 and NCT01598311) [57, 58]. No clinical trials in children are available.

Tigecycline

Tigecycline belongs to the glycylcycline class and mainly inhibits protein synthesis. This intravenous antibiotic has bacteriostatic activity against *C. difficile* and, although not registered for being used to treat CDI, its use is approved for skin and intra-abdominal infections and community-acquired pneumonia [59, 60]. In a retrospective, observational cohort study with participants treated with intravenous tigecycline (tigecycline group) and with both oral vancomycin and intravenous metronidazole (standard therapy group), Gergely Szabo *et al.* [61] reported higher clinical cure rates for the tigecycline group (75.6%) in comparison with the standard therapy group (53.3%). Despite the promising results, no clinical trials are available on the treatment with this antibiotic, to our knowledge.

OTHER PROMISING ANTI-INFECTIVE STRATEGIES FOR THE PREVENTION/TREATMENT OF CDI

In addition to developing new antibiotics, alternative natural therapies are also being developed for use in the treatment of initial episodes of CDI. New studies have increasingly been carried out to reduce the risk of recurrences and severe forms of CDI. Although several biological therapies are being developed, many of them still have no conclusive studies. Except for monoclonal antibodies and probiotics, no other clinical studies with emerging therapies in children are available.

Vaccines

Toxoid vaccines have been developed against *C. difficile* toxins with transformed toxin structures and producing antitoxin A and B antibodies. The development of recurrent CDI is inversely correlated with antibody levels against *C. difficile* toxins from the patient, which means that vaccines against *C. difficile* toxins may have the potential to prevent both initial and recurrent CDI [62]. No vaccine is so far available on the market, and even though the demonstrated efficacy in several studies, many challenges are involved in the production of a toxoid vaccine, such as manipulation of toxin, decontamination of spores, and quantitative regulation of each toxin produced by the same *C. difficile* strain [63]. Toxoid vaccines are generally well-tolerated, and adverse events are mainly related to local reactions in the injection site [64].

A few vaccines are still in the initial phase of development. The safety and the immunogenic response of a *C. difficile* investigational vaccine based on the F2 antigen (GSK2904545A) and developed by GlaxoSmithKline are being investigated in an ongoing phase I clinical trial (NCT04026009) with healthy adults aged 18-45 and 50-70 years old. In an open-label phase I clinical trial, vaccine VLA84/IC84 (Valneva Austria GmbH), a recombinant fusion protein comprising portions of *C. difficile* toxins A and B, was demonstrated to be safe, well-tolerated, and highly immunogenic in volunteer adults aged ≥ 18 to < 65 and ≥ 65 [65]. It was also immunogenic in all doses and formulations tested, independently of the volunteers' age (50-64 and ≥ 65 years old), in a phase II clinical trial [66]. However, a phase III trial has not yet started.

A formalin-inactivated toxoid-based vaccine developed by Sanofi Pasteur, effective in phase I and phase II clinical trials [67, 68], was discontinued after analysis of a phase III clinical trial by the Independent Data Monitoring Committee, which concluded that the main objective of the study had a low probability of being achieved (NCT01887912). Two ongoing phases III clinical trials are currently investigating a *C. difficile* vaccine (Pfizer) in healthy adults 50 years of age and older (NCT03918629 and NCT03090191), but data on their efficacy is not yet available.

The *C. difficile* immunogenic lipoprotein CD0873, firstly identified in 2014 [68], was recently related to the prevention of *C. difficile* long-term gut colonization and a strong secretory IgA immune response [69]. This suggests that CD0873 may be one of the next components against which to develop a new vaccine for CDI.prevention.

Monoclonal Antibodies

Monoclonal antibodies are capable of neutralizing *C. difficile* toxins, and thus preventing their cytotoxic effects in the intestinal mucosa, and assisting restoration of the normal gut microbiota [62]. Actoxumab and bezlotoxumab are human monoclonal antibodies developed by Merck against *C. difficile* toxins A and B, respectively. Even though the *in vitro* ability to neutralize the toxin A, thus preventing its association with target cells resulting in the inhibition of the first step of the intoxication cascade [70], the effectiveness of actoxumab was not observed when administrated (10 mg/kg intravenously) in a phase III clinical trial to study the efficacy of an individual infusion of actoxumab, bezlotoxumab and actoxumab plus bezlotoxumab in patients receiving antibiotic therapy for CDI (NCT01241552, MODIFY I). This trial was, therefore, discontinued after a planned interim analysis [71]. In the same phase III study and in a second randomized, double-blind, and placebo-controlled phase III study (NCT01513239, MODIFY II), also the treatment of a single dose of bezlotoxumab and bezlotoxumab in combination with actoxumab was administrated intravenously, and similar efficacies were found with significantly lower recurrence rates than placebo groups. Therefore, only bezlotoxumab has FDA approval since 2016 to prevent CDI recurrence in adult patients being treated with antibacterial drugs for CDI and at high-risk for recurrence [71]. The efficacy of bezlotoxumab in children aged 1 to < 18 years old, diagnosed with CDI, and receiving antibacterial drug treatment is also being investigated in an ongoing phase III clinical trial (NCT03182907) with a completion date estimated for May 2022.

Nevertheless, more studies are still required. On the one hand, although it was not statistically significant, eight patients experienced heart failure, which suggests that bezlotoxumab may not be recommended for patients with cardiovascular diseases [71]. On the other hand, the high cost of the treatment remains a problem.

Non-Toxigenic Spores

Based on evidence that colonization with non-toxigenic *C. difficile* could prevent CDI by toxigenic strains showed improvements in humans with recurrent CDI [72]. Villano *et al.* evaluated an oral suspension of spores of non-toxigenic *C. difficile* strain M3 (VP20621) administered to adult volunteers following antibiotic treatment in phase I clinical trial (ViroPharma) [73]. The researchers found that VP20621 was well tolerated, and high rates of colonization were observed. In phase II clinical trial (NCT01259726), VP20621 was well tolerated at all investigated doses and allowed a significant reduction of CDI recurrence [74]. Besides the potential of VP20621 as a biotherapeutic strategy to prevent the recurrence of CDI, other trials are necessary, and the possibility of the acquisition of genes encoding toxins by non-toxigenic strains should not be excluded.

Bacteriophages

Bacteriophages, or phages, are viruses that uniquely infect bacteria and have the ability to modulate bacterial communities [75]. The inactivity of bacteriophages in the extracellular environment ensures their safety for animals and plants [76], but their specificity to bacterial species and/or strains is an ambiguous issue compared to the broad-spectrum antibiotics. If, on the one hand, this specificity is an advantage regarding side effects on the intestinal microbiome, which can lead to CDI episodes, on the other, it may be less effecacious on infections that are caused by more than one type of *C. difficile* strain [77]. The use of "phage cocktails" consisting of multiple infective phages against various bacterial strains could be a strategy to overcome phage specificity [77].

Although accepted the use of temperate phages, *i.e.*, phages undergoing lysogenic infections without bacterial cell death, it is recommended to use strictly lytic phages for therapy, *i.e.*, virulent phages that kill the target bacteria [78]. This is a concern since none of the 25 complete genomes of *C. difficile* phages currently available are strictly lytic [79]. However, the effort in isolating strictly lytic phages can also be overcome with phage combinations that help reduce the effect of lysogeny and possible resistances. Some *in vitro* and *in vivo* studies have explored the potential of phage combinations as a therapeutic approach for CDI. Nale *et al.* [80] studied the impact of an optimized 4-phage cocktail on a *C. difficile* ribotype 014/020 clinical strain and observed complete *C. difficile* eradication (6 log reduction) after 24h following prophylactic or remedial regimens as well as the non-affected viability of commensal bacteria. In another study, the same authors reported that a cocktail of temperate phages significantly reduced *C. difficile* growth *in vitro* and proliferation *in vivo* using a hamster model [81]. Nonetheless, the *in vitro* activity of phages is not always directly translated into *in vivo* animal models, so a larger number of studies is needed before assays in animal models and human tests are undertaken. In an attempt to simulate the phage-bacteria interaction in the colon environment, Shan *et al.* [82] used the human colon tumorigenic cell line HT-29 and observed a more effective reduction of *C. difficile* by the phage phiCDHS1 in the presence of HT-29 cells than in bacterial culture. The authors believe that human cell lines are ideal for understanding the biology of phages *in vivo* and to predict clinical responses, which would be very useful in providing data to be applied in human trials [82].

As important as understanding how phages affect *C. difficile* in the colon environment, it is equally necessary to know how phages affect the gut microbiome. In a recent cohort study, the success of fecal microbiota transplantation to treat recurrent CDI was positively correlated with the relative abundance of bacteriophage in the donor [83]. Zuo *et al.* [84] also noted that the colonization of *Caudovirales* bacteriophages from donors following FMT helped in the treatment response in adult recipients with CDI in a preliminary study belonging to an ongoing clinical trial (NCT02570477).

Undoubtedly, future investigations to address the safety concerns of phage therapy and clinical data supported by credible clinical trials are needed to demonstrate the efficacy of phage therapy for the treatment of CDI.

Probiotics

Probiotics are defined as "live microorganisms which when administered in adequate amounts confer a health benefit on the host", according to the Food and Agriculture Organization of the United Nations (FAO) and the World Health Organization (WHO) Expert Committee [85]. The most common probiotics tested against *C. difficile* belong to *Lactobacillus* and *Bifidobacterium* genera. However, the probiotic yeast *Saccharomyces boulardii* and some probiotic combinations have also been suggested (Fig. **2**), despite their clinical role as preventive and/or treatment of CDI remaining undefined.

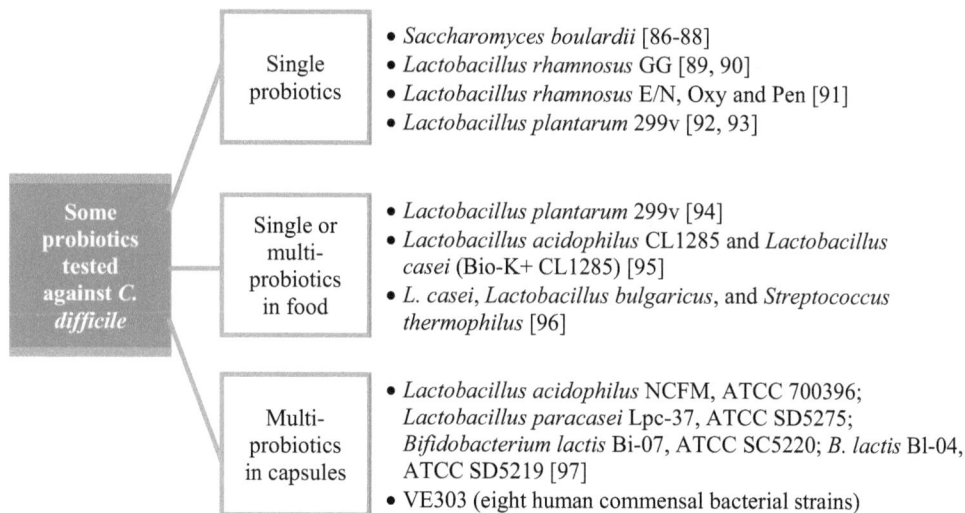

Some probiotics tested against *C. difficile*

Single probiotics
- *Saccharomyces boulardii* [86-88]
- *Lactobacillus rhamnosus* GG [89, 90]
- *Lactobacillus rhamnosus* E/N, Oxy and Pen [91]
- *Lactobacillus plantarum* 299v [92, 93]

Single or multi-probiotics in food
- *Lactobacillus plantarum* 299v [94]
- *Lactobacillus acidophilus* CL1285 and *Lactobacillus casei* (Bio-K+ CL1285) [95]
- *L. casei*, *Lactobacillus bulgaricus*, and *Streptococcus thermophilus* [96]

Multi-probiotics in capsules
- *Lactobacillus acidophilus* NCFM, ATCC 700396; *Lactobacillus paracasei* Lpc-37, ATCC SD5275; *Bifidobacterium lactis* Bi-07, ATCC SC5220; *B. lactis* Bl-04, ATCC SD5219 [97]
- VE303 (eight human commensal bacterial strains)

Fig. (2). Examples of probiotic formulations tested against *C. difficile*.

In a double-blind, randomized, placebo-controlled, and parallel-group intervention study, McFarland *et al.* [86] combined *S. boulardii* with vancomycin or metronidazole for four weeks in the treatment of patients with active CDI and found that although there were no benefits for patients with initial episodes, the efficacy of *S. boulardii* was significant for those with recurrent CDI. In contrast, Surawicz *et al.* [87] reported that the difference in the CDI recurrence rate in patients treated with vancomycin and *S. boulardii* was not significant compared to patients treated with placebo. In a more recent multi-center, double-masked, randomized, placebo-controlled phase III clinical trial (NCT01143272), which was finalized after a masked independent interim analysis, no evidence for an effect of *S. boulardii* in preventing *C. difficile*-associated diarrhea was found for hospitalized adult patients without particular risk factors [88].

Lactobacillus rhamnosus GG (ATCC 53103) was firstly isolated from the intestinal tract of a healthy human in 1983 (Patent US4839281A), and its efficacy in the treatment of adult patients with a multiple recurrent CDI was observed in an open-label study published in 1987 [89]. Conversely, in two randomized and double-blind studies conducted years later, *L. rhamnosus* GG was ineffective in preventing CDI in hospitalized adults receiving antibiotics [90]. In another double-blind, randomized, and placebo-controlled clinical trial, other *L. rhamnosus* strains (E/N, Oxy, and Pen) were considered safe, well-tolerated, and effective in the prevention of antibiotic-associated diarrhea caused by *C. difficile* in children [91].

The role of *Lactobacillus plantarum* 299v (DSM 9843) in preventing recurrent CDI has also been extensively explored. In a double-blind and placebo-controlled trial, although recurrence of CDI occurred more frequently in patients receiving metronidazole and placebo than those receiving metronidazole combined with *L. plantarum* 299v, the small size of the sample has not allowed drawing conclusions about *L. plantarum* 299v effectiveness [92]. It is expected that an ongoing multi-center, randomized, placebo-controlled, parallel-group, and two-sided superiority trial (ANZCTR12617000783325) will conclude about the restoration of the gut microbiota in adult patients admitted to an intensive care unit and receiving *L. plantarum* 299v in addition to the usual treatment [93]. In another study with intensive care unit patients receiving a fermented oatmeal gruel containing *L. plantarum* 299v, the colonization of *C. difficile* was reduced [94].

Other foods combining probiotic and prebiotic formulations have been studied as promising strategies to prevent or treat CDI. In a randomized, double-blind, and placebo-controlled trial, fermented milk consumption with *Lactobacillus acidophilus* CL1285 and *Lactobacillus casei* (Bio-K+ CL1285) by hospitalized patients was effective in preventing CDI and allowed a lower median hospitalization duration [95]. Similarly, in another double-blind, randomized, and placebo-controlled study (N0016106821), a probiotic drink combining *L. casei*,

Lactobacillus bulgaricus, and *Streptococcus thermophilus* reduced the incidence of CDI in adult patients [96].

Recently, a multi-strain capsule (*L. acidophilus* NCFM, ATCC 700396; *Lactobacillus paracasei* Lpc-37, ATCC SD5275; *Bifidobacterium lactis* Bi-07, ATCC SC5220; *B. lactis* Bl-04, ATCC SD5219) was administered daily for four weeks in a randomized and phase II clinical trial (NCT01680874) to determine the probiotics' ability to reduce diarrhea duration in patients with an initial episode of mild to moderate CDI. The administration of this capsule reduced the diarrhea duration, but the small number of patients necessitates replication of the study in a larger population in order to draw more accurate conclusions [97]. Another multi-strain capsule (VE303) with eight human commensal bacterial strains is being developed by Vedanta Biosciences, Inc. for the prevention of recurrent CDI. After a successful phase I study, a randomized and double-blind phase II study (NCT03788434) is being conducted to evaluate the safety, toleration, and efficacy of VE303 in adults with one or more CDI recurrences.

It is empirical that the restoration of an altered intestinal microbiome will help prevent an opportunistic infection as CDI and, eventually, would help in its elimination. Despite all the reported benefits of probiotics in preventing and/or treating CDI, additional larger prospective and comparative studies are needed due to some contradictory results. Long-term safety and side-effects of probiotics, especially among immunocompromised individuals and critically ill patients, should also be considered in further studies.

Bacteriocins

Bacteriocins are antimicrobial peptides synthesized ribosomally during translation by Gram-negative and -positive bacteria that have bactericidal or bacteriostatic activity against closely related species (small-spectrum) or across genera (broad-spectrum) [98]. The most common bacteriocins are those produced by Gram-positive bacteria and are categorized into different classes according to their structure and mode of action. Briefly, class I comprises low-molecular-weight (<5 kDa) peptides and is divided into class Ia (lantibiotics), Ib (labyrinthopeptins), and Ic (sactibiotics). Class II and class III comprise small (<10 kDa) heat-stable peptides and large (>10 kDa) heat-labile proteins, respectively [99]. The potential of several bacteriocins specifically targeting *C. difficile* has been reported, but few have been tested in clinical trials (Table 1).

Lantibiotics

Nisin is the best-characterized lantibiotic produced by *Lactococcus* and *Streptococcus* species with a broad-spectrum activity against several Gram-positive bacteria, including *C. difficile* [100]. Several naturally-occurring variants of nisin and their activity *in vitro* against *C. difficile* have been reported, such as nisin A produced by *Lactococcus lactis* [101] or nisin Z by *Lactococcus lactis* NIZO 22186 [102]. Optimized bioengineered forms of nisin have also been created, as in the case of nisin V (M21V), which exhibited antimicrobial activity *in vitro* against *C. difficile* strain ribotype 027 with two-fold increased activity (Minimum Inhibitory Concentration (MIC) of 4.19 µg/ml) than the parent nisin A (MIC of 8.38 µg/ml) [103]. Although promising results against *C. difficile*, the broad-spectrum activity of nisin against several Gram-positive bacteria would cause a similar effect on the gut microbiome. Le Lay *et al.* [104] demonstrated that commercial nisin reduced both *C. difficile* and Gram-positive bacteria of the gut, but that bacterial balance was restored within 24 h following treatment.

Lacticin 3147 is produced by *L. lactis* subsp. *lactis* DPC3147 and, like nisin, has a broad-spectrum activity against several bacterial species, including *C. difficile*. Rea *et al.* [105] demonstrated that the addition of lacticin 3147, vancomycin or metronidazole to a human distal colon model had a significant impact on the numbers of *C. difficile* but also on the composition of the gut microbiome, which means that further optimization of the lacticin 3147 is necessary before considering its use as a treatment option of CDI.

Actagardine is produced by *Actinoplanes garbadinensis* and *Actinoplanes liguriae* with a broad-spectrum activity against Gram-positive bacteria [106]. From the naturally-occurring variant actagardine A, produced by *A. garbadinensis* ATCC 31049, Boakes *et al.* [107] generated several mutants, including V15F. This mutant demonstrates a remarkably improved activity against four *C. difficile* strains compared to the wild-type actagardine A and vancomycin. NVB302 is a semi-synthetic variant of deoxyactagardine B, a structural analog of actagardine A produced by *A. liguriae* NCIMB41362 [108]. The response to this peptide and vancomycin was compared by Crowther *et al.* [109] in an *in vitro* human gut model, and in addition to NVB302 being non-inferior to vancomycin in the treatment of simulated CDI, its effect was less harmful against the group of *Bacteroides fragilis* from the gut microbiome. After the promising pre-clinical results, a phase I clinical trial (ISRCTN40071144) was conducted for NVB302, but no results were published to date.

Mutacin 1140 is produced by *Streptococcus mutans*, which holds the largest number of lantibiotic variants described to date [110, 111]. Selected mutacin 1140 variants OG253 and OG716 have been studied for the treatment of CDI. Kers *et al.* [110] reported OG253 as one of the most stable variants with low toxicity against HepG2 cells and high MIC against *C. difficile in vitro*, as well as with the potential to inhibit CDI relapse in OG253-treated hamsters. Pre-clinical evaluation of the maximum tolerated dose and toxicokinetics of OG253 capsules was performed by Rajeshkumar *et al.* [112]. Despite the need for further pre-clinical studies under more prolonged exposure, the oral administration of three doses of OG253 capsules (6.75, 27, and 108 mg/day) to Wistar Han rats in a one-day dose-escalation study, followed by a seven-day repeated-dose toxicokinetics study, was demonstrated to be well tolerated and did not alter the weight of organs nor affect the hematology, coagulation, clinical biochemistry parameters and urine pH compared to placebo capsules. Furthermore, the authors recommended 425.7 mg/kg/day as the maximum tolerated dose of OG253 [112]. The potential of mutacin-variant OG716 has also been demonstrated *in vivo* in Golden Syrian hamster models of CDI [111, 113]. Kers *et al.* [111] found that oral OG716 conferred 100% survival of hamsters and with no relapse at three weeks post-infection, and Pulse *et al.* [113] reported no observable toxicity, side-effects or intestinal motility after oral administration of the maximum possible dose of OG716 ($\geq$1.918 mg/kg/day). All reported data support the importance of the clinical development of mutacin 1140 variants as a potential strategy on recurrent CDI.

Other lantibiotics such as mersacidin [114] produced by *Bacillus* sp. strain HIL Y-85/54728, microbisporicin NAI-107 [115] produced by *Microbispora* sp. strain ATCC-PTA-5024, and formicin [116] produced by *Bacillus paralicheniformis* APC 1576 have also demonstrated antimicrobial activity against a few strains of *C. difficile*, but no pre-clinical studies have been published.

Thiopeptides

LFF571 is a semi-synthetic derivative of GE2270, a natural thiopeptide bacteriocin isolated in 1991 from a *Planobispora rosea* strain [117], that inhibits Gram-positive bacteria, including *C. difficile*, but has lower antimicrobial activity against *Lactobacillus* and *Bifidobacterium* species [118]. LFF571 proved to be safe and well-tolerated in a randomized, double-blind, placebo-controlled phase I clinical trial in healthy volunteers [119]. A multi-center, randomized, evaluator-blind, and active-controlled phase II clinical trial (NCT01232595) assessing the safety, efficacy, and pharmacokinetics of multiple daily dosing of oral LFF571 and vancomycin in adult patients with primary episodes or first relapses of moderate CDI was completed in 2015 [120, 121]. Regarding pharmacokinetics parameters, drug concentrations were measured in serum and fecal samples and, in contrast with vancomycin, LFF571 exhibited low serum levels (41.7 ng/ml) and high fecal concentrations (107 and 12,900 µg/g) demonstrating its retention in the gastrointestinal lumen [120]. The clinical cure rate for LFF571 was not inferior to vancomycin, and also, the recurrence rates of CDI were lower. Even though LFF571 was demonstrated to be safe and well-tolerated, the incidence of adverse events was higher for LFF571 [121], revealing the need for further studies. Additionally, subinhibitory concentrations of LFF571 can decrease the production of toxins A and B from *C. difficile in vitro* more rapidly than colony formation inhibition [122]. So, the studies performed so far demonstrated that LFF571 represents an excellent candidate for CDI treatment, although further investigation is needed.

Sactibiotics

Thuricin CD is produced by *Bacillus thuringiensis* DPC 6431 with a narrow-spectrum activity against *C. difficile*, including the hypervirulent *C. difficile* BI/NAP1/027 strain [123]. Antimicrobial activity *in vitro* was confirmed in a distal colon model, and thuricin CD was demonstrated to be equally effective against *C. difficile* and to have a low impact on the composition of the gut microbiome [105]. Mathur *et al.* [106] evaluated the anti-*C. difficile* activity of thuricin CD, vancomycin, metronidazole, ramoplanin, and actagardine and found that thuricin CD did not produce antagonistic effects when combined with the tested antibiotics and, alone or in combination with ramoplanin, demonstrated significant activity against *C. difficile*. Later, Mathur *et al.* [124] evaluated the efficacy of thuricin CD, tigecycline, vancomycin, teicoplanin, rifampicin, and nitazoxanide against *C. difficile* biofilms and planktonic cells and found that except in the combination of thuricin CD-nitazoxanide, all the other combinations were highly effective against two *C. difficile* biofilm producer strains. With the presented results, it seems that thuricin CD could play an essential role in reducing CDI recurrence rates, mainly when caused by *C. difficile* biofilm-forming strains. Moreover, the narrow-spectrum activity reported and the minimal impact on the gut microbiome composition reveals that thuricin CD is a promising therapeutic strategy against CDI, despite no clinical trials have been conducted yet.

Table 1. Examples of bacteriocins with anti-*C. difficile* activity.

Bacteriocin	Producer	References
Lantibiotic		
Nisin A (nisin natural variant)	*Lactococcus lactis* strains	[101]
Nisin Z (nisin natural variant)	*Lactococcus lactis* NIZO22186	[102]
Nisin V M21V (nisin A bioengineered variant)	*Lactococcus lactis* NZ9800	[103]
Lacticin 3147	*Lactococcus lactis* subsp. *lactis* DPC3147	[105]
V15F (actagardine A bioengineered variant)	*Actinoplanes garbadinensis* ATCC 31049	[107]

(Table 1) cont.....

Bacteriocin	Producer	References
NVB302 (deoxyactagardine B semi-synthetic variant)	*Actinoplanes liguriae* NCIMB41362 *liguriae*	[108, 109]
OG253 (mutacin 1140 bioengineered variant)	*Streptococcus mutans* strains	[110, 112]
OG716 (mutacin 1140 bioengineered variant)	*Streptococcus mutans* strains	[111, 113]
Mersacidin	*Bacillus* sp. strain HIL Y-85/54728	[114]
NAI-107	*Microbispora* sp. strain ATCC-PTA-5024	[115]
Formicin	*Bacillus paralicheniformis* APC 1576	[116]
Thiopeptide		
LFF571 (GE2210 semi-synthetic variant)	*Planobispora rosea* strain	[118 - 122]
Sactibiotic		
Thuricin CD	*Bacillus thuringiensis* DPC 6431	[105, 106, 123, 124]

Synthetic Polymers

Synthetic polymers with direct therapeutic effects could have a wide range of applications, binding agent polymers being the most explored [125]. A promising strategy against the CDI was tolevamer, a soluble polymer that binds and neutralizes *C. difficile* toxins A and B *in vitro* [126]. However, two randomized, double-blinded, and controlled phase III trials (NCT00106509 and NCT00196794) were conducted to compare the efficacy of tolevamer with vancomycin and metronidazole for the treatment of CDI; it was observed that the clinical responses to tolevamer were inferior to those of both antibiotics [127]. Even so, due to the ability to neutralize *C. difficile* toxins, the use of this polymer should be considered as a possible adjuvant therapy.

Polymers with antimicrobial activity and the potential to combat CDI, nylon-3-polymers, were discovered by Liu *et al.* (US20170000818A1 patent) [128]. These polymers (poly-β-peptides) have been investigated as artificial imitators of host-defense peptides but with advantages due to their easier synthesis than peptides themselves and their resistance to proteolysis. The authors found that nylon-3-polymers can influence *C. difficile* spore survival, germination, or the subsequent growth into vegetative form. This means that nylon-3-polymers may have a crucial role in controlling infection by *C. difficile* and preventing the production of their toxins, which makes these polymers appealing to clinical applications. Nevertheless, as far as we are aware, no more studies are ongoing for nylon-3-polymers.

CONCLUDING REMARKS

Clostridioides difficile infection causes diarrheal disease with potentially fatal complications. The current treatments available are limited, including vancomycin, metronidazole, and fidaxomicin, and their increased failure in both cure rates and recurrences of CDI remains a problem. Therefore, alternative solutions are a top priority, and consequently, the number of clinical studies has increased considerably in recent years.

Fecal microbiota transplantation is emerging as a therapeutic option in treating CDI recurrences, but although beneficial effects have been demonstrated, further research is still necessary to increase knowledge about its long-term safety. Conventional therapeutics continue to focus on major pharmaceutical industries, with a large number of sponsored clinical trials. This includes the development of new antibiotics such as nitazoxanide, ramoplanin, or tigecycline. However, most have failed in clinical trials due to their inferiority to the antibiotics already used, and only the new antibiotics CRS3123 and ridinilazole are awaiting the results of the ongoing phase II and III clinical trials, respectively. Also, the development of vaccines and monoclonal antibodies has been attempted by the pharmaceutical industry. The monoclonal antibody bezlotoxumab, whilst still being investigated for its effectiveness in children with diagnosed CDI, was approved by the FDA in 2016 as an adjuvant therapy to prevent CDI in adults treated with antibiotics for CDI and with a high-risk of recurrence. However, there is not yet a vaccine available for CDI prevention (initial or recurrent), and only one is being studied in an ongoing phase III clinical trial (Pfizer).

It is evident that the promising strategies in the treatment of CDI necessarily involve maintaining the integrity of the intestinal microbiome, in contrast to the imbalance mediated by the traditional first-line antibiotics. Many studies have been carried out to discover natural therapies, such as the case of non-pathogenic bacterial species, probiotics, bacteriophages, or peptides produced by bacteria with remarkable activities against *C. difficile*. The specificity of bacteriophages and the use of phage cocktails that infect a large number of *C. difficile* strains without affecting the gut microbiome is a promising strategy in the treatment of CDI and should be further investigated in the future. A considerable number of bacteriocins have shown *in vitro* activity against *C. difficile*, but *in vivo* studies are still scarce. Two bacteriocins with only a limited effect on the gut microbiome, the lantibiotic NVB302 completed a phase I study (results not yet published), and a thiopeptide bacteriocin LFF571 proved to be safe, well-tolerated, and not inferior to vancomycin in a phase II clinical trial. Despite interesting strategies *per se*, the use of bacteriocins as adjuvant therapeutic agents can be an interesting approach to increase their efficiency. The use of probiotics, alone or in combination, as prophylaxis for CDI prevention is promising but still controversial. The best choice appears to be the combination of different species and/or strains, which implies further optimization in terms of specificity, appropriate mixtures, optimum doses, safety, or effectiveness.

Infection control measures are admittedly essential anti-infective strategies against CDI, which is proven by its decreased incidence in most European countries from 2004 to 2015 [129], in Canada from 2009 to 2015 [130], and in the USA from 2015 to 2019 [131]. Unfortunately, this epidemiological trend is not observed worldwide, and with this review, the authors have attempted to summarize most of the strategies under development to fight CDI.

Microbiological challenges are rapidly increasing worldwide. Based on the vast information that already exists, it is necessary to invest in more research to study these promising new anti-infective agents against CDI to overcome this enormous problem as soon as possible.

CONSENT FOR PUBLICATION

Not Applicable.

CONFLICT OF INTEREST

The author declares no conflict of interest, financial or otherwise.

ACKNOWLEDGEMENTS

The authors are grateful to Dr. Paul Gibbs for the English edition of this manuscript. We would also like to thank the scientific collaboration under the *Fundação para a Ciência e a Tecnologia* (FCT) project UID/Multi/50016/2019. Financial support for author J. Barbosa was provided by a post- doctoral fellowship SFRH/BPD/113303/2015 (FCT).

REFERENCES

[1]   Lawson PA, Citron DM, Tyrrell KL, Finegold SM. Reclassification of *Clostridium difficile* as *Clostridioides difficile* (Hall and O'Toole 1935) Prévot 1938. Anaerobe 2016; 40: 95-9.
        [http://dx.doi.org/10.1016/j.anaerobe.2016.06.008] [PMID: 27370902]

[2]   Hall EC, O'Toole E. Intestinal flora in new-born infants with a description of a new pathogenic anaerobe, *Bacillus difficilis*. Am J Dis Child 1935; 49(2): 390-02.
        [http://dx.doi.org/10.1001/archpedi.1935.019700020105010]

[3]   Bartlett JG, Moon N, Chang TW, Taylor N, Onderdonk AB. Role of *Clostridium difficile* in antibiotic-associated pseudomembranous colitis. Gastroenterology 1978; 75(5): 778-82.
        [http://dx.doi.org/10.1016/0016-5085(78)90457-2] [PMID: 700321]

[4]   Kachrimanidou M, Malisiovas N. *Clostridium difficile* infection: a comprehensive review. Crit Rev Microbiol 2011; 37(3): 178-87.

[http://dx.doi.org/10.3109/1040841X.2011.556598] [PMID: 21609252]

[5]   Rupnik M, Wilcox MH, Gerding DN. *Clostridium difficile* infection: new developments in epidemiology and pathogenesis. Nat Rev Microbiol 2009; 7(7): 526-36.
[http://dx.doi.org/10.1038/nrmicro2164] [PMID: 19528959]

[6]   Bartlett JG. Narrative review: the new epidemic of *Clostridium difficile*-associated enteric disease. Ann Intern Med 2006; 145(10): 758-64.
[http://dx.doi.org/10.7326/0003-4819-145-10-200611210-00008] [PMID: 17116920]

[7]   Gerding DN, Johnson S, Peterson LR, Mulligan ME, Silva J Jr. *Clostridium difficile*-associated diarrhea and colitis. Infect Control Hosp Epidemiol 1995; 16(8): 459-77.
[http://dx.doi.org/10.2307/30141083] [PMID: 7594392]

[8]   Zar FA, Bakkanagari SR, Moorthi KMLST, Davis MB. A comparison of vancomycin and metronidazole for the treatment of *Clostridium difficile*-associated diarrhea, stratified by disease severity. Clin Infect Dis 2007; 45(3): 302-7.
[http://dx.doi.org/10.1086/519265] [PMID: 17599306]

[9]   Hopkins RJ, Wilson RB. Treatment of recurrent *Clostridium difficile* colitis: a narrative review. Gastroenterol Rep (Oxf) 2018; 6(1): 21-8.
[http://dx.doi.org/10.1093/gastro/gox041] [PMID: 29479439]

[10] Rupnik M, Dupuy B, Fairweather NF, *et al*. Revised nomenclature of *Clostridium difficile* toxins and associated genes. J Med Microbiol 2005; 54(Pt 2): 113-7.
[http://dx.doi.org/10.1099/jmm.0.45810-0] [PMID: 15673503]

[11] von Eichel-Streiber C, Boquet P, Sauerborn M, Thelestam M. Large clostridial cytotoxins--a family of glycosyltransferases modifying small GTP-binding proteins. Trends Microbiol 1996; 4(10): 375-82.
[http://dx.doi.org/10.1016/0966-842X(96)10061-5] [PMID: 8899962]

[12] Terhes G, Urbán E, Sóki J, Hamid KA, Nagy E. Community-acquired *Clostridium difficile* diarrhea caused by binary toxin, toxin A, and toxin B gene-positive isolates in Hungary. J Clin Microbiol 2004; 42(9): 4316-8.
[http://dx.doi.org/10.1128/JCM.42.9.4316-4318.2004] [PMID: 15365032]

[13] Carter GP, Lyras D, Allen DL, *et al*. Binary toxin production in *Clostridium difficile* is regulated by CdtR, a LytTR family response regulator. J Bacteriol 2007; 189(20): 7290-301.
[http://dx.doi.org/10.1128/JB.00731-07] [PMID: 17693517]

[14] Barbut F, Decré D, Lalande V, *et al*. Clinical features of *Clostridium difficile*-associated diarrhoea due to binary toxin (actin-specific ADP-ribosyltransferase)-producing strains. J Med Microbiol 2005; 54(Pt 2): 181-5.
[http://dx.doi.org/10.1099/jmm.0.45804-0] [PMID: 15673514]

[15] Bauer MP, Veenendaal D, Verhoef L, Bloembergen P, van Dissel JT, Kuijper EJ. Clinical and microbiological characteristics of community-onset *Clostridium difficile* infection in The Netherlands. Clin Microbiol Infect 2009; 15(12): 1087-92.
[http://dx.doi.org/10.1111/j.1469-0691.2009.02853.x] [PMID: 19624512]

[16] Gupta A, Khanna S. Community-acquired *Clostridium difficile* infection: an increasing public health threat. Infect Drug Resist 2014; 7: 63-72.
[PMID: 24669194]

[17] McDonald LC, Gerding DN, Johnson S, *et al*. Clinical practice guidelines for *Clostridium difficile* infection in adults and children: 2017 update by the Infectious Diseases Society of America (IDSA) and Society for Healthcare Epidemiology of America (SHEA). Clin Infect Dis 2018; 66(7): e1-e48.
[http://dx.doi.org/10.1093/cid/cix1085] [PMID: 29462280]

[18] Lo Vecchio A, Lancella L, Tagliabue C, *et al*. *Clostridium difficile* infection in children: epidemiology and risk of recurrence in a low-prevalence country. Eur J Clin Microbiol Infect Dis 2017; 36(1): 177-85.

[http://dx.doi.org/10.1007/s10096-016-2793-7] [PMID: 27696233]

[19]   Malmqvist L, Ullberg M, Hed Myrberg I, Nilsson A. *Clostridium difficile* infection in children: epidemiology and trend in a Swedish tertiary care hospital. Pediatr Infect Dis J 2019; 38(12): 1208-13.
[http://dx.doi.org/10.1097/INF.0000000000002480] [PMID: 31738336]

[20]   McDonald LC, Killgore GE, Thompson A, *et al.* An epidemic, toxin gene-variant strain of *Clostridium difficile.* N Engl J Med 2005; 353(23): 2433-41.
[http://dx.doi.org/10.1056/NEJMoa051590] [PMID: 16322603]

[21]   Akerlund T, Persson I, Unemo M, *et al.* Increased sporulation rate of epidemic *Clostridium difficile* Type 027/NAP1. J Clin Microbiol 2008; 46(4): 1530-3.
[http://dx.doi.org/10.1128/JCM.01964-07] [PMID: 18287318]

[22]   Warny M, Pepin J, Fang A, *et al.* Toxin production by an emerging strain of *Clostridium difficile* associated with outbreaks of severe disease in North America and Europe. Lancet 2005; 366(9491): 1079-84.
[http://dx.doi.org/10.1016/S0140-6736(05)67420-X] [PMID: 16182895]

[23]   Debast SB, Bauer MP, Kuijper EJ. European society of clinical microbiology and infectious diseases: update of the treatment guidance document for clostridium difficile infection. Clin Microbiol Infect 2014; 20(2) (Suppl. 2): 1-26.
[http://dx.doi.org/10.1111/1469-0691.12418] [PMID: 24118601]

[24]   Credito KL, Appelbaum PC. Activity of OPT-80, a novel macrocycle, compared with those of eight other agents against selected anaerobic species. Antimicrob Agents Chemother 2004; 48(11): 4430-4.
[http://dx.doi.org/10.1128/AAC.48.11.4430-4434.2004] [PMID: 15504874]

[25]   Tannock GW, Munro K, Taylor C, *et al.* A new macrocyclic antibiotic, fidaxomicin (OPT-80), causes less alteration to the bowel microbiota of *Clostridium difficile*-infected patients than does vancomycin. Microbiology (Reading) 2010; 156(Pt 11): 3354-9.
[http://dx.doi.org/10.1099/mic.0.042010-0] [PMID: 20724385]

[26]   Louie TJ, Miller MA, Mullane KM, *et al.* OPT-80-003 Clinical Study Group. Fidaxomicin *versus* vancomycin for *Clostridium difficile* infection. N Engl J Med 2011; 364(5): 422-31.
[http://dx.doi.org/10.1056/NEJMoa0910812] [PMID: 21288078]

[27]   Nerandzic MM, Mullane K, Miller MA, Babakhani F, Donskey CJ. Reduced acquisition and overgrowth of vancomycin-resistant enterococci and *Candida* species in patients treated with fidaxomicin *versus* vancomycin for *Clostridium difficile* infection. Clin Infect Dis 2012; 55(2) (Suppl. 2): S121-6.
[http://dx.doi.org/10.1093/cid/cis440] [PMID: 22752860]

[28]   Choi HK, Kim KH, Lee SH, Lee SJ. Risk factors for recurrence of *Clostridium difficile* infection: effect of vancomycin-resistant enterococci colonization. J Korean Med Sci 2011; 26(7): 859-64.
[http://dx.doi.org/10.3346/jkms.2011.26.7.859] [PMID: 21738336]

[29]   Yamamoto T, Abe K, Anjiki H, Ishii T, Kuyama Y. Metronidazole-induced neurotoxicity developed in liver cirrhosis. J Clin Med Res 2012; 4(4): 295-8.
[PMID: 22870180]

[30]   Wolf J, Kalocsai K, Fortuny C, *et al.* Safety and efficacy of fidaxomicin and vancomycin in children and adolescents with *Clostridioides* (*Clostridium*) *difficile* infection: a phase 3, multi-center, randomized, single-blind clinical trial (SUNSHINE). Clin Infect Dis 2019; ciz1149.
[http://dx.doi.org/10.1093/cid/ciz1149]

[31]   Song JH, Kim YS. Recurrent *Clostridium difficile* infection: risk factors, treatment, and prevention. Gut Liver 2019; 13(1): 16-24.
[http://dx.doi.org/10.5009/gnl18071] [PMID: 30400734]

[32]   Martin LT, Vincent S, Gillian S, Moore K, Ratermann D, Droege CA. Pharmacologic approach to management of *Clostridium difficile* infection. Crit Care Nurs Q 2019; 42(1): 2-11.

[http://dx.doi.org/10.1097/CNQ.0000000000000232] [PMID: 30507659]

[33]   Cammarota G, Masucci L, Ianiro G, *et al.* Randomised clinical trial: faecal microbiota transplantation by colonoscopy *vs.* vancomycin for the treatment of recurrent *Clostridium difficile* infection. Aliment Pharmacol Ther 2015; 41(9): 835-43.
[http://dx.doi.org/10.1111/apt.13144] [PMID: 25728808]

[34]   Vindigni SM, Surawicz CM. Fecal microbiota transplantation. Gastroenterol Clin North Am 2017; 46(1): 171-85.
[http://dx.doi.org/10.1016/j.gtc.2016.09.012] [PMID: 28164849]

[35]   Kelly CR, Khoruts A, Staley C, *et al.* Effect of fecal microbiota transplantation on recurrence in multiply recurrent *Clostridium difficile* infection: a randomized trial. Ann Intern Med 2016; 165(9): 609-16.
[http://dx.doi.org/10.7326/M16-0271] [PMID: 27547925]

[36]   Russell G, Kaplan J, Ferraro M, Michelow IC. Fecal bacteriotherapy for relapsing *Clostridium difficile* infection in a child: a proposed treatment protocol. Pediatrics 2010; 126(1): e239-42.
[http://dx.doi.org/10.1542/peds.2009-3363] [PMID: 20547640]

[37]   Walia R, Garg S, Song Y, *et al.* Efficacy of fecal microbiota transplantation in 2 children with recurrent *Clostridium difficile* infection and its impact on their growth and gut microbiome. J Pediatr Gastroenterol Nutr 2014; 59(5): 565-70.
[http://dx.doi.org/10.1097/MPG.0000000000000495] [PMID: 25023578]

[38]   Nicholson MR, Mitchell PD, Alexander E, *et al.* Efficacy of fecal microbiota transplantation for *Clostridium difficile* infection in children. Clin Gastroenterol Hepatol 2020; 18(3): 612-619.e1.
[http://dx.doi.org/10.1016/j.cgh.2019.04.037] [PMID: 31009795]

[39]   Locher HH, Seiler P, Chen X, *et al. In vitro* and *in vivo* antibacterial evaluation of cadazolid, a new antibiotic for treatment of *Clostridium difficile* infections. Antimicrob Agents Chemother 2014; 58(2): 892-900.
[http://dx.doi.org/10.1128/AAC.01830-13] [PMID: 24277020]

[40]   Gerding DN, Hecht DW, Louie T, *et al.* Susceptibility of *Clostridium difficile* isolates from a Phase 2 clinical trial of cadazolid and vancomycin in *C. difficile* infection. J Antimicrob Chemother 2016; 71(1): 213-9.
[http://dx.doi.org/10.1093/jac/dkv300] [PMID: 26433782]

[41]   Louie T, Nord CE, Talbot GH, *et al.* Multicenter, double-blind, randomized, phase 2 study evaluating the novel antibiotic cadazolid in patients with *Clostridium difficile* infection. Antimicrob Agents Chemother 2015; 59(10): 6266-73.
[http://dx.doi.org/10.1128/AAC.00504-15] [PMID: 26248357]

[42]   Gerding DN, Cornely OA, Grill S, *et al.* Cadazolid for the treatment of *Clostridium difficile* infection: results of two double-blind, placebo-controlled, non-inferiority, randomised phase 3 trials. Lancet Infect Dis 2019; 19(3): 265-74.
[http://dx.doi.org/10.1016/S1473-3099(18)30614-5] [PMID: 30709665]

[43]   Ochsner UA, Bell SJ, O'Leary AL, *et al.* Inhibitory effect of REP3123 on toxin and spore formation in *Clostridium difficile*, and *in vivo* efficacy in a hamster gastrointestinal infection model. J Antimicrob Chemother 2009; 63(5): 964-71.
[http://dx.doi.org/10.1093/jac/dkp042] [PMID: 19251726]

[44]   Nayak SU, Griffiss JM, Blumer J, *et al.* Safety, tolerability, systemic exposure, and metabolism of CRS3123, a methionyl-tRNA synthetase inhibitor developed for treatment of *Clostridium difficile*, in a phase 1 study. Antimicrob Agents Chemother 2017; 61(8): e02760-16.
[http://dx.doi.org/10.1128/AAC.02760-16] [PMID: 28584140]

[45]   Dubreuil L, Houcke I, Mouton Y, Rossignol JF. *In vitro* evaluation of activities of nitazoxanide and tizoxanide against anaerobes and aerobic organisms. Antimicrob Agents Chemother 1996; 40(10): 2266-70.

[http://dx.doi.org/10.1128/AAC.40.10.2266] [PMID: 8891127]

[46]   Musher DM, Logan N, Hamill RJ, *et al.* Nitazoxanide for the treatment of *Clostridium difficile* colitis. Clin Infect Dis 2006; 43(4): 421-7.
[http://dx.doi.org/10.1086/506351] [PMID: 16838229]

[47]   Musher DM, Logan N, Bressler AM, Johnson DP, Rossignol JF. Nitazoxanide *versus* vancomycin in *Clostridium difficile* infection: a randomized, double-blind study. Clin Infect Dis 2009; 48(4): e41-6.
[http://dx.doi.org/10.1086/596552] [PMID: 19133801]

[48]   Trubiano JA, Cheng AC, Korman TM, *et al.* Australasian Society of Infectious Diseases updated guidelines for the management of *Clostridium difficile* infection in adults and children in Australia and New Zealand. Intern Med J 2016; 46(4): 479-93.
[http://dx.doi.org/10.1111/imj.13027] [PMID: 27062204]

[49]   Peláez T, Alcalá L, Alonso R, *et al. In vitro* activity of ramoplanin against *Clostridium difficile*, including strains with reduced susceptibility to vancomycin or with resistance to metronidazole. Antimicrob Agents Chemother 2005; 49(3): 1157-9.
[http://dx.doi.org/10.1128/AAC.49.3.1157-1159.2005] [PMID: 15728918]

[50]   Freeman J, Baines SD, Jabes D, Wilcox MH. Comparison of the efficacy of ramoplanin and vancomycin in both *in vitro* and *in vivo* models of clindamycin-induced *Clostridium difficile* infection. J Antimicrob Chemother 2005; 56(4): 717-25.
[http://dx.doi.org/10.1093/jac/dki321] [PMID: 16143709]

[51]   Pullman J, Prieto J, Leach TS. Ramoplanin *vs.* vancomycin in the treatment of *Clostridium difficile* diarrhea: a phase II study (Abstracts). Proceedings of the 44th Interscience conference on antimicrobial agents and chemotherapy.

[52]   Wong MT, Kauffman CA, Standiford HC, *et al.* Ramoplanin VRE2 Clinival Stury Group. Effective suppression of vancomycin-resistant *Enterococcus* species in asymptomatic gastrointestinal carriers by a novel glycolipodepsipeptide, ramoplanin. Clin Infect Dis 2001; 33(9): 1476-82.
[http://dx.doi.org/10.1086/322687] [PMID: 11588692]

[53]   Corbett D, Wise A, Birchall S, *et al. In vitro* susceptibility of *Clostridium difficile* to SMT19969 and comparators, as well as the killing kinetics and post-antibiotic effects of SMT19969 and comparators against *C. difficile.* J Antimicrob Chemother 2015; 70(6): 1751-6.
[http://dx.doi.org/10.1093/jac/dkv006] [PMID: 25652750]

[54]   Vickers RJ, Tillotson GS, Nathan R, *et al.* CoDIFy study group. Efficacy and safety of ridinilazole compared with vancomycin for the treatment of *Clostridium difficile* infection: a phase 2, randomised, double-blind, active-controlled, non-inferiority study. Lancet Infect Dis 2017; 17(7): 735-44.
[http://dx.doi.org/10.1016/S1473-3099(17)30235-9] [PMID: 28461207]

[55]   Alam MZ, Wu X, Mascio C, Chesnel L, Hurdle JG. Mode of action and bactericidal properties of surotomycin against growing and nongrowing *Clostridium difficile*. Antimicrob Agents Chemother 2015; 59(9): 5165-70.
[http://dx.doi.org/10.1128/AAC.01087-15] [PMID: 26055381]

[56]   Lee CH, Patino H, Stevens C, *et al.* Surotomycin *versus* vancomycin for *clostridium difficile* infection: Phase 2, randomized, controlled, double-blind, non-inferiority, multicentre trial. J Antimicrob Chemother 2016; 71(10): 2964-71.
[http://dx.doi.org/10.1093/jac/dkw246] [PMID: 27432604]

[57]   Boix V, Fedorak RN, Mullane KM, *et al.* Primary outcomes from a phase 3, randomized, double-blind, active-controlled trial of surotomycin in subjects with *Clostridium difficile* infection. Open Forum Infect Dis 2017; 4(1): ofw275.
[http://dx.doi.org/10.1093/ofid/ofw275] [PMID: 28480267]

[58]   Daley P, Louie T, Lutz JE, *et al.* Surotomycin *versus* vancomycin in adults with *Clostridium difficile* infection: primary clinical outcomes from the second pivotal, randomized, double-blind, Phase 3 trial. J Antimicrob Chemother 2017; 72(12): 3462-70.

[http://dx.doi.org/10.1093/jac/dkx299] [PMID: 28961905]

[59]   Babinchak T, Ellis-Grosse E, Dartois N, Rose GM, Loh E. Tigecycline 301 Study Group; Tigecycline 306 Study Group. The efficacy and safety of tigecycline for the treatment of complicated intra-abdominal infections: analysis of pooled clinical trial data. Clin Infect Dis 2005; 41(5) (Suppl. 5): S354-67.
[http://dx.doi.org/10.1086/431676] [PMID: 16080073]

[60]   Nathwani D. Tigecycline: clinical evidence and formulary positioning. Int J Antimicrob Agents 2005; 25(3): 185-92.
[http://dx.doi.org/10.1016/j.ijantimicag.2004.11.006] [PMID: 15737510]

[61]   Gergely Szabo B, Kadar B, Szidonia Lenart K, *et al.* Use of intravenous tigecycline in patients with severe *Clostridium difficile* infection: a retrospective observational cohort study. Clin Microbiol Infect 2016; 22(12): 990-5.
[http://dx.doi.org/10.1016/j.cmi.2016.08.017] [PMID: 27599690]

[62]   Dieterle MG, Rao K, Young VB. Novel therapies and preventative strategies for primary and recurrent *Clostridium difficile* infections. Ann N Y Acad Sci 2019; 1435(1): 110-38.
[http://dx.doi.org/10.1111/nyas.13958] [PMID: 30238983]

[63]   Roshan N, Hammer KA, Riley TV. Non-conventional antimicrobial and alternative therapies for the treatment of *Clostridium difficile* infection. Anaerobe 2018; 49: 103-11.
[http://dx.doi.org/10.1016/j.anaerobe.2018.01.003] [PMID: 29309845]

[64]   Henderson M, Bragg A, Fahim G, Shah M, Hermes-DeSantis ER. A review of the safety and efficacy of vaccines as prophylaxis for *Clostridium difficile* infections. Vaccines (Basel) 2017; 5(3): 25.
[http://dx.doi.org/10.3390/vaccines5030025] [PMID: 28869502]

[65]   Bézay N, Ayad A, Dubischar K, *et al.* Safety, immunogenicity and dose response of VLA84, a new vaccine candidate against *Clostridium difficile*, in healthy volunteers. Vaccine 2016; 34(23): 2585-92.
[http://dx.doi.org/10.1016/j.vaccine.2016.03.098] [PMID: 27079932]

[66]   Bézay N, Hochreiter R, Jelinek T, Kadlecek V, Kiermayr S, Dubischar K. A phase 2, dose-confirmation immunogenicity and safety study of Vla84, a *Clostridium difficile* vaccine candidate, in adults aged 50 years and older (abstract). Proceedings of the ASM Microbe. Boston, Massachusetts, USA. 2016.

[67]   de Bruyn G, Saleh J, Workman D, *et al.* H-030-012 Clinical Investigator Study Team. Defining the optimal formulation and schedule of a candidate toxoid vaccine against *Clostridium difficile* infection: A randomized Phase 2 clinical trial. Vaccine 2016; 34(19): 2170-8.
[http://dx.doi.org/10.1016/j.vaccine.2016.03.028] [PMID: 27013431]

[68]   Kovacs-Simon A, Leuzzi R, Kasendra M, Minton N, Titball RW, Michell SL. Lipoprotein CD0873 is a novel adhesin of *Clostridium difficile*. J Infect Dis 2014; 210(2): 274-84.
[http://dx.doi.org/10.1093/infdis/jiu070] [PMID: 24482399]

[69]   Bradshaw WJ, Bruxelle JF, Kovacs-Simon A, *et al.* Molecular features of lipoprotein CD0873: A potential vaccine against the human pathogen *Clostridioides difficile*. J Biol Chem 2019; 294(43): 15850-61.
[http://dx.doi.org/10.1074/jbc.RA119.010120] [PMID: 31420448]

[70]   Hernandez LD, Kroh HK, Hsieh E, *et al.* Epitopes and mechanism of action of the *Clostridium difficile* toxin A-neutralizing antibody actoxumab. J Mol Biol 2017; 429(7): 1030-44.
[http://dx.doi.org/10.1016/j.jmb.2017.02.010] [PMID: 28232034]

[71]   Wilcox MH, Gerding DN, Poxton IR, *et al.* MODIFY I and MODIFY II Investigators. Bezlotoxumab for prevention of recurrent *Clostridium difficile* infection. N Engl J Med 2017; 376(4): 305-17.
[http://dx.doi.org/10.1056/NEJMoa1602615] [PMID: 28121498]

[72]   Seal D, Borriello SP, Barclay F, Welch A, Piper M, Bonnycastle M. Treatment of relapsing *Clostridium difficile* diarrhoea by administration of a non-toxigenic strain. Eur J Clin Microbiol 1987;

6(1): 51-3.
[http://dx.doi.org/10.1007/BF02097191] [PMID: 3569251]

[73] Villano SA, Seiberling M, Tatarowicz W, Monnot-Chase E, Gerding DN. Evaluation of an oral suspension of VP20621, spores of nontoxigenic *Clostridium difficile* strain M3, in healthy subjects. Antimicrob Agents Chemother 2012; 56(10): 5224-9.
[http://dx.doi.org/10.1128/AAC.00913-12] [PMID: 22850511]

[74] Gerding DN, Meyer T, Lee C, *et al.* Administration of spores of nontoxigenic *Clostridium difficile* strain M3 for prevention of recurrent *C. difficile* infection: a randomized clinical trial. JAMA 2015; 313(17): 1719-27.
[http://dx.doi.org/10.1001/jama.2015.3725] [PMID: 25942722]

[75] Sabino J, Hirten RP, Colombel J-F. Review article: bacteriophages in gastroenterology-from biology to clinical applications. Aliment Pharmacol Ther 2020; 51(1): 53-63.
[http://dx.doi.org/10.1111/apt.15557] [PMID: 31696976]

[76] Atterbury RJ. Bacteriophage biocontrol in animals and meat products. Microb Biotechnol 2009; 2(6): 601-12.
[http://dx.doi.org/10.1111/j.1751-7915.2009.00089.x] [PMID: 21255295]

[77] Lin DM, Koskella B, Lin HC. Phage therapy: An alternative to antibiotics in the age of multi-drug resistance. World J Gastrointest Pharmacol Ther 2017; 8(3): 162-73.
[http://dx.doi.org/10.4292/wjgpt.v8.i3.162] [PMID: 28828194]

[78] Moelling K, Broecker F, Willy C. A wake-up call: we need phage therapy now. Viruses 2018; 10(12): 688.
[http://dx.doi.org/10.3390/v10120688] [PMID: 30563034]

[79] Li T, Zhang Y, Dong K, *et al.* Isolation and characterization of the novel phage JD032 and global transcriptomic response during JD032 infection of *Clostridioides difficile* ribotype 078. mSystems 2020; 5(3): e00017-20.
[http://dx.doi.org/10.1128/mSystems.00017-20] [PMID: 32371470]

[80] Nale JY, Redgwell TA, Millard A, Clokie MRJ. Efficacy of an optimised bacteriophage cocktail to clear *Clostridium difficile* in a batch fermentation model. Antibiotics (Basel) 2018; 7(1): 13.
[http://dx.doi.org/10.3390/antibiotics7010013] [PMID: 29438355]

[81] Nale JY, Spencer J, Hargreaves KR, *et al.* Bacteriophage combinations significantly reduce *Clostridium difficile* growth *in vitro* and proliferation *in vivo*. Antimicrob Agents Chemother 2015; 60(2): 968-81.
[http://dx.doi.org/10.1128/AAC.01774-15] [PMID: 26643348]

[82] Shan J, Ramachandran A, Thanki AM, Vukusic FBI, Barylski J, Clokie MRJ. Bacteriophages are more virulent to bacteria with human cells than they are in bacterial culture; insights from HT-29 cells. Sci Rep 2018; 8(1): 5091.
[http://dx.doi.org/10.1038/s41598-018-23418-y] [PMID: 29572482]

[83] Park H, Laffin MR, Jovel J, *et al.* The success of fecal microbial transplantation in *Clostridium difficile* infection correlates with bacteriophage relative abundance in the donor: a retrospective cohort study. Gut Microbes 2019; 10(6): 676-87.
[http://dx.doi.org/10.1080/19490976.2019.1586037] [PMID: 30866714]

[84] Zuo T, Wong SH, Lam K, *et al.* Bacteriophage transfer during faecal microbiota transplantation in *Clostridium difficile* infection is associated with treatment outcome. Gut 2018; 67(4): 634-43.
[PMID: 28539351]

[85] FAO/WHO. Evaluation of health and nutritional properties of powder milk and live lactic acid bacteria. Food and Agriculture Organization of the United Nations and World Health Organization expert consultation report 2001.

[86] McFarland LV, Surawicz CM, Greenberg RN, *et al.* A randomized placebo-controlled trial of

Saccharomyces boulardii in combination with standard antibiotics for *Clostridium difficile* disease. JAMA 1994; 271(24): 1913-8.
[http://dx.doi.org/10.1001/jama.1994.03510480037031] [PMID: 8201735]

[87]   Surawicz CM, McFarland LV, Greenberg RN, *et al.* The search for a better treatment for recurrent *Clostridium difficile* disease: use of high-dose vancomycin combined with *Saccharomyces boulardii*. Clin Infect Dis 2000; 31(4): 1012-7.
[http://dx.doi.org/10.1086/318130] [PMID: 11049785]

[88]   Ehrhardt S, Guo N, Hinz R, *et al. Saccharomyces boulardii* to prevent antibiotic-associated diarrhea: a randomized, double-masked, placebo-controlled trial. Open Forum Infect Dis 2016; 3(1): ofw011.
[http://dx.doi.org/10.1093/ofid/ofw011] [PMID: 26973849]

[89]   Gorbach SL, Chang TW, Goldin B. Successful treatment of relapsing *Clostridium difficile* colitis with *Lactobacillus* GG. Lancet 1987; 2(8574): 1519.
[http://dx.doi.org/10.1016/S0140-6736(87)92646-8] [PMID: 2892070]

[90]   Miller M, Florencio S, Eastmond J, Reynolds S. Results of 2 prospective randomized studies of *Lactobacillus* GG to prevent *C. difficile* infection in hospitalized adults receiving antibiotics (abstract K-4200). Proceedings of the 48th Interscience Conference on Antimicrobial Agents and Chemotherapy. Washington, DC, USA. 2008; pp. 578-9.

[91]   Ruszczyński M, Radzikowski A, Szajewska H. Clinical trial: effectiveness of *Lactobacillus rhamnosus* (strains E/N, Oxy and Pen) in the prevention of antibiotic-associated diarrhoea in children. Aliment Pharmacol Ther 2008; 28(1): 154-61.
[http://dx.doi.org/10.1111/j.1365-2036.2008.03714.x] [PMID: 18410562]

[92]   Wullt M, Hagslätt ML, Odenholt I. *Lactobacillus plantarum* 299v for the treatment of recurrent *Clostridium difficile*-associated diarrhoea: a double-blind, placebo-controlled trial. Scand J Infect Dis 2003; 35(6-7): 365-7.
[http://dx.doi.org/10.1080/003655403310010985] [PMID: 12953945]

[93]   Litton E, Anstey M, Broadhurst D, *et al.* Study protocol for the safety and efficacy of probiotic therapy on days alive and out of hospital in adult ICU patients: the multicentre, randomised, placebo-controlled Restoration Of gut microflora in Critical Illness Trial (ROCIT). BMJ Open 2020; 10(6): e035930.
[http://dx.doi.org/10.1136/bmjopen-2019-035930] [PMID: 32565465]

[94]   Klarin B, Wullt M, Palmquist I, Molin G, Larsson A, Jeppsson B. *Lactobacillus plantarum* 299v reduces colonisation of *Clostridium difficile* in critically ill patients treated with antibiotics. Acta Anaesthesiol Scand 2008; 52(8): 1096-102.
[http://dx.doi.org/10.1111/j.1399-6576.2008.01748.x] [PMID: 18840110]

[95]   Beausoleil M, Fortier N, Guénette S, *et al.* Effect of a fermented milk combining *Lactobacillus acidophilus* Cl1285 and *Lactobacillus casei* in the prevention of antibiotic-associated diarrhea: a randomized, double-blind, placebo-controlled trial. Can J Gastroenterol 2007; 21(11): 732-6.
[http://dx.doi.org/10.1155/2007/720205] [PMID: 18026577]

[96]   Hickson M, D'Souza AL, Muthu N, *et al.* Use of probiotic *Lactobacillus* preparation to prevent diarrhoea associated with antibiotics: randomised double blind placebo controlled trial. BMJ 2007; 335(7610): 80.
[http://dx.doi.org/10.1136/bmj.39231.599815.55] [PMID: 17604300]

[97]   Barker AK, Duster M, Valentine S, *et al.* A randomized controlled trial of probiotics for *Clostridium difficile* infection in adults (PICO). J Antimicrob Chemother 2017; 72(11): 3177-80.
[http://dx.doi.org/10.1093/jac/dkx254] [PMID: 28961980]

[98]   Cotter PD, Ross RP, Hill C. Bacteriocins - a viable alternative to antibiotics? Nat Rev Microbiol 2013; 11(2): 95-105.
[http://dx.doi.org/10.1038/nrmicro2937] [PMID: 23268227]

[99]   Kumariya R, Garsa AK, Rajput YS, Sood SK, Akhtar N, Patel S. Bacteriocins: Classification,

synthesis, mechanism of action and resistance development in food spoilage causing bacteria. Microb Pathog 2019; 128: 171-7.
[http://dx.doi.org/10.1016/j.micpath.2019.01.002] [PMID: 30610901]

[100] Kerr KG, Copley RM, Wilcoy MH. Activity of nisin against *Clostridium difficile*. Lancet 1997; 349(9057): 1026-7.
[http://dx.doi.org/10.1016/S0140-6736(05)62927-3] [PMID: 9100650]

[101] Bartoloni A, Mantella A, Goldstein BP, *et al. In-vitro* activity of nisin against clinical isolates of *Clostridium difficile*. J Chemother 2004; 16(2): 119-21.
[http://dx.doi.org/10.1179/joc.2004.16.2.119] [PMID: 15216943]

[102] Lay CL, Dridi L, Bergeron MG, Ouellette M, Fliss IL. Nisin is an effective inhibitor of *Clostridium difficile* vegetative cells and spore germination. J Med Microbiol 2016; 65(2): 169-75.
[http://dx.doi.org/10.1099/jmm.0.000202] [PMID: 26555543]

[103] Field D, Quigley L, O'Connor PM, *et al.* Studies with bioengineered Nisin peptides highlight the broad-spectrum potency of Nisin V. Microb Biotechnol 2010; 3(4): 473-86.
[http://dx.doi.org/10.1111/j.1751-7915.2010.00184.x] [PMID: 21255345]

[104] Le Lay C, Fernandez B, Hammami R, Ouellette M, Fliss I. On *Lactococcus lactis* UL719 competitivity and nisin (Nisaplin(®)) capacity to inhibit *Clostridium difficile* in a model of human colon. Front Microbiol 2015; 6: 1020.
[http://dx.doi.org/10.3389/fmicb.2015.01020] [PMID: 26441942]

[105] Rea MC, Dobson A, O'Sullivan O, *et al.* Effect of broad- and narrow-spectrum antimicrobials on *Clostridium difficile* and microbial diversity in a model of the distal colon. Proc Natl Acad Sci USA 2011; 108(1) (Suppl. 1): 4639-44.
[http://dx.doi.org/10.1073/pnas.1001224107] [PMID: 20616009]

[106] Mathur H, O'Connor PM, Hill C, Cotter PD, Ross RP. Analysis of anti-*Clostridium difficile* activity of thuricin CD, vancomycin, metronidazole, ramoplanin, and actagardine, both singly and in paired combinations. Antimicrob Agents Chemother 2013; 57(6): 2882-6.
[http://dx.doi.org/10.1128/AAC.00261-13] [PMID: 23571539]

[107] Boakes S, Ayala T, Herman M, Appleyard AN, Dawson MJ, Cortés J. Generation of an actagardine A variant library through saturation mutagenesis. Appl Microbiol Biotechnol 2012; 95(6): 1509-17.
[http://dx.doi.org/10.1007/s00253-012-4041-0] [PMID: 22526797]

[108] Boakes S, Appleyard AN, Cortés J, Dawson MJ. Organization of the biosynthetic genes encoding deoxyactagardine B (DAB), a new lantibiotic produced by *Actinoplanes liguriae* NCIMB41362. J Antibiot (Tokyo) 2010; 63(7): 351-8.
[http://dx.doi.org/10.1038/ja.2010.48] [PMID: 20520597]

[109] Crowther GS, Baines SD, Todhunter SL, Freeman J, Chilton CH, Wilcox MH. Evaluation of NVB302 *versus* vancomycin activity in an *in vitro* human gut model of *Clostridium difficile* infection. J Antimicrob Chemother 2013; 68(1): 168-76.
[http://dx.doi.org/10.1093/jac/dks359] [PMID: 22966180]

[110] Kers JA, Sharp RE, Defusco AW, *et al.* Mutacin 1140 lantibiotic variants are efficacious against *Clostridium difficile* infection. Front Microbiol 2018; 9(9): 415. a
[http://dx.doi.org/10.3389/fmicb.2018.00415] [PMID: 29615987]

[111] Kers JA, DeFusco AW, Park JH, *et al.* OG716: Designing a fit-for-purpose lantibiotic for the treatment of *Clostridium difficile* infections. PLoS One 2018; 13(6): e0197467. b
[http://dx.doi.org/10.1371/journal.pone.0197467] [PMID: 29894469]

[112] Rajeshkumar NV, Kers JA, Moncrief S, Defusco AW, Park JH, Handfield M. Preclinical evaluation of the maximum tolerated dose and toxicokinetics of enteric-coated lantibiotic OG253 capsules. Toxicol Appl Pharmacol 2019; 374: 32-40.
[http://dx.doi.org/10.1016/j.taap.2019.04.019] [PMID: 31034929]

[113]   Pulse ME, Weiss WJ, Kers JA, DeFusco AW, Park JH, Handfield M. Pharmacological, toxicological, and dose range assessment of OG716, a novel lantibiotic for the treatment of *Clostridium difficile*-associated infection. Antimicrob Agents Chemother 2019; 63(4): e01904-18.
[http://dx.doi.org/10.1128/AAC.01904-18] [PMID: 30670434]

[114]   Niu WW, Neu HC. Activity of mersacidin, a novel peptide, compared with that of vancomycin, teicoplanin, and daptomycin. Antimicrob Agents Chemother 1991; 35(5): 998-1000.
[http://dx.doi.org/10.1128/AAC.35.5.998] [PMID: 1649577]

[115]   Castiglione F, Lazzarini A, Carrano L, *et al.* Determining the structure and mode of action of microbisporicin, a potent lantibiotic active against multiresistant pathogens. Chem Biol 2008; 15(1): 22-31.
[http://dx.doi.org/10.1016/j.chembiol.2007.11.009] [PMID: 18215770]

[116]   Collins FWJ, O'Connor PM, O'Sullivan O, Rea MC, Hill C, Ross RP. Formicin - a novel broad-spectrum two-component lantibiotic produced by *Bacillus paralicheniformis* APC 1576. MicroSoc 2016; 162(9)

[117]   Selva E, Beretta G, Montanini N, *et al.* Antibiotic GE2270 a: a novel inhibitor of bacterial protein synthesis. I. Isolation and characterization. J Antibiot (Tokyo) 1991; 44(7): 693-701.
[http://dx.doi.org/10.7164/antibiotics.44.693] [PMID: 1908853]

[118]   Citron DM, Tyrrell KL, Merriam CV, Goldstein EJ. Comparative *in vitro* activities of LFF571 against *Clostridium difficile* and 630 other intestinal strains of aerobic and anaerobic bacteria. Antimicrob Agents Chemother 2012; 56(5): 2493-503.
[http://dx.doi.org/10.1128/AAC.06305-11] [PMID: 22290948]

[119]   Ting LS, Praestgaard J, Grunenberg N, Yang JC, Leeds JA, Pertel P. A first-in-human, randomized, double-blind, placebo-controlled, single- and multiple-ascending oral dose study to assess the safety and tolerability of LFF571 in healthy volunteers. Antimicrob Agents Chemother 2012; 56(11): 5946-51.
[http://dx.doi.org/10.1128/AAC.00867-12] [PMID: 22964250]

[120]   Bhansali SG, Mullane K, Ting LS, *et al.* Pharmacokinetics of LFF571 and vancomycin in patients with moderate *Clostridium difficile* infections. Antimicrob Agents Chemother 2015; 59(3): 1441-5.
[http://dx.doi.org/10.1128/AAC.04252-14] [PMID: 25534724]

[121]   Mullane K, Lee C, Bressler A, *et al.* Multicenter, randomized clinical trial to compare the safety and efficacy of LFF571 and vancomycin for *Clostridium difficile* infections. Antimicrob Agents Chemother 2015; 59(3): 1435-40.
[http://dx.doi.org/10.1128/AAC.04251-14] [PMID: 25534727]

[122]   Sachdeva M, Leeds JA. Subinhibitory concentrations of LFF571 reduce toxin production by *Clostridium difficile.* Antimicrob Agents Chemother 2015; 59(2): 1252-7.
[http://dx.doi.org/10.1128/AAC.04436-14] [PMID: 25512411]

[123]   Rea MC, Sit CS, Clayton E, *et al.* Thuricin CD, a posttranslationally modified bacteriocin with a narrow spectrum of activity against *Clostridium difficile.* Proc Natl Acad Sci USA 2010; 107(20): 9352-7.
[http://dx.doi.org/10.1073/pnas.0913554107] [PMID: 20435915]

[124]   Mathur H, Rea MC, Cotter PD, Hill C, Ross RP. The efficacy of thuricin CD, tigecycline, vancomycin, teicoplanin, rifampicin and nitazoxanide, independently and in paired combinations against *Clostridium difficile* biofilms and planktonic cells. Gut Pathog 2016; 8: 20.
[http://dx.doi.org/10.1186/s13099-016-0102-8] [PMID: 27257437]

[125]   Connor EF, Lees I, Maclean D. Polymers as drugs—Advances in therapeutic applications of polymer binding agents. J Polym Sci A Polym Chem 2017; 55: 3146-57.
[http://dx.doi.org/10.1002/pola.28703]

[126]   Kurtz CB, Cannon EP, Brezzani A, *et al.* GT160-246, a toxin binding polymer for treatment of

Clostridium difficile colitis. Antimicrob Agents Chemother 2001; 45(8): 2340-7.
[http://dx.doi.org/10.1128/AAC.45.8.2340-2347.2001] [PMID: 11451694]

[127]   Johnson S, Louie TJ, Gerding DN, *et al.* Polymer Alternative for CDI Treatment (PACT) investigators. Vancomycin, metronidazole, or tolevamer for *Clostridium difficile* infection: results from two multinational, randomized, controlled trials. Clin Infect Dis 2014; 59(3): 345-54.
[http://dx.doi.org/10.1093/cid/ciu313] [PMID: 24799326]

[128]   Liu R, Suárez JM, Weisblum B, Gellman SH, McBride SM. Synthetic polymers active against *Clostridium difficile* vegetative cell growth and spore outgrowth. J Am Chem Soc 2014; 136(41): 14498-504.
[http://dx.doi.org/10.1021/ja506798e] [PMID: 25279431]

[129]   Ho J, Wong SH, Doddangoudar VC, Boost MV, Tse G, Ip M. Regional differences in temporal incidence of *Clostridium difficile* infection: a systematic review and meta-analysis. Am J Infect Control 2020; 48(1): 89-94.
[http://dx.doi.org/10.1016/j.ajic.2019.07.005] [PMID: 31387772]

[130]   Katz KC, Golding GR, Choi KB, *et al.* Canadian Nosocomial Infection Surveillance Program. The evolving epidemiology of *Clostridium difficile* infection in Canadian hospitals during a postepidemic period (2009-2015). CMAJ 2018; 190(25): E758-65.
[http://dx.doi.org/10.1503/cmaj.180013] [PMID: 29941432]

[131]   Centers for Disease Control and Prevention. US Department of Health and Human Services
https://arpsp.cdc.gov/profile/infections/CDI

CHAPTER 4

Anti-*Toxoplasma* Drug Discovery and Natural Products: a Brief Overview

Jhony Anacleto-Santos[1,2], Perla Y López-Camacho[2], Elisa Vega-Ávila[3], Ricardo Mondragón-Flores[4], Elba Carrasco-Ramírez[5] and **Norma Rivera-Fernández[5,*]**

[1] *Doctorado en Ciencias Biológicas y de la Salud. Universidad Autónoma Metropolitana, Mexico*

[2] *Departamento de Ciencias Naturales, Universidad Autónoma Metropolitana, Unidad Cuajimalpa, 05348 CDMX, Mexico*

[3] *Departamento de Ciencias de la Salud, Universidad Autónoma Metropolitana, Unidad Iztapalapa, 09340 CDMX, Mexico*

[4] *Departamento de Bioquímica, Centro de Investigación y Estudios Avanzados del Instituto Politécnico Nacional (CINVESTAV-IPN), Zacatenco 07360, CDMX, Mexico*

[5] *Departamento de Microbiología y Parasitología, Facultad de Medicina, Universidad Nacional Autónoma de México (UNAM). Coyoacán, 04510 CDMX, Mexico*

Abstract: *Toxoplasma gondii*, an apicomplexan protozoan that is considered an opportunistic parasite of medical and veterinary interest, causes toxoplasmosis, which may be asymptomatic in the immunocompetent host, while fatal in the immunocompromised patient. A combination of pyrimethamine-sulfadiazine is the treatment of choice; nevertheless, these two drugs produce severe side effects and are only effective against acute toxoplasmosis, hence, less toxic novel compounds with anti-toxoplasma activity are greatly needed. Natural compounds seem to be a promising source to identify lead compounds against *T. gondii*. In this review, the *in vitro* and *in vivo* activities of extracts, fractions, and isolated compounds obtained from different plants are described. In addition, some biological and pathological generalities of the parasite are reviewed as well. Data were obtained from a bibliographic search throughout digital faculty libraries, Google Scholar search engine, Science Direct, SciELO databases, and the National Center for Biotechnology Information (NCBI). Founded records include the evaluation of 58 extracts, 14 fractions, and 7 compounds belonging to 53 species of 33 plant families. Predominant studies were made on *in vivo* RH *T. gondii* tachyzoites strain and very few in *in vivo* models of both acute and chronic toxoplasmosis. Research of natural compounds against *T. gondii* deserves more attention in order to identify novel drugs useful in the control of toxoplasmosis.

---

[*] **Corresponding author Norma Rivera-Fernández:** Departamento de Microbiología y Parasitología, Facultad de Medicina, UNAM CDMX 01820, Mexico; E-mail: normariv@unam.mx

Keywords: Anti-*Toxoplasma*, Bradyzoite, Medicinal plants, Natural products, Plant extracts, *Toxoplasma gondii*, Toxoplasmicide, Tissue cysts, Tachyzoite, IC^{50}.

INTRODUCTION

Based on their multiple active metabolites with therapeutic properties, natural products have been used for thousands of years to treat human malaises. Nowadays, natural products are still used by approximately four out of five people worldwide as teas, tinctures, extracts or alimentary supplements. Throughout the modern history of mankind, natural compounds have been used to design novel drugs with different uses, such as antimicrobials, antipyretics, and anti-inflammatories. The main antimalarials commonly used around the world were synthesized from natural compounds. Chloroquine was developed from quinine, an alkaloid obtained from the Cinchona stem bark; artesunate and artemether were derived from an artemisinin compound found in *Artemisia annua* leaves [1 - 4]. Based on the fact that plants have been a successful source of antimalarials, the search for novel anti-*Toxoplasma* compounds from natural products has recently increased [5 - 9]. In this review, the *in vitro* and *in vivo* activities of extracts, fractions, and isolated compounds obtained from different plants are described. To better understand the need to develop new treatments, some general aspects regarding *Toxoplasma gondii* biology and toxoplasmosis control, need to be understood. These aspects are herein briefly described.

T. gondii is an apicomplexan parasite of medical and veterinary importance and perhaps the most successful and ubiquitous protozoan. During its life cycle, the parasite develops different infectious phases, as tachyzoites, tissue cysts containing bradyzoites and oocysts [10, 11]. Crescent shape intracellular tachyzoites are responsible for the acute phase of the infection and can disseminate through blood or lymph to invade all nucleated cells [11] (Fig. **1**). Proinflammatory cytokines (mostly interferon-gamma) participate in a not well characterized molecular pathway that allows tachyzoites transformation into bradyzoites and tissue cysts, that are most frequently observed in the immunocompetent host brain [12, 13] (Fig. **2**). The slow proliferation of bradyzoites contained in tissue cysts causes chronic infections and can eventually interconvert into tachyzoites [10]. Oocysts, which are the resistance phase of the parasite, are found in cat feces and become infectious after approximately five days in the environment withstanding extreme conditions for several months [10]. *T. gondii* is considered a food-borne parasite as it can be transmitted by the consumption of raw or poorly cooked meat infected with tissue cysts or by the ingesta of food and water contaminated with infectious oocysts [11]. Tachyzoite infections are rarely reported to occur after drinking unpasteurized milk or by laboratory accidents. Infection can also occur by transplacental route, transplantation of infected organs or by blood transfusion. *T. gondii* life cycle can be observed in Fig. (**3**). Cats´ hygiene habits, as well as parasites´ life cycle, make direct transmission difficult. *T. gondii* infection goes clinically unnoticed in the immunocompetent host whereas it can be lethal in immunocompromised patients [14, 15]. Unspecific clinical manifestations, such as lymphadenitis, fever, odynophagia, cephalea, myalgias, and in rare cases hepatosplenomegaly, pulmonary or cardiac symptoms are reported in the acute phase. Toxoplasmic encephalitis can lead to death in AIDS patients, while the parasite can cause abortion or malformations in the newborn [13, 16 - 18].

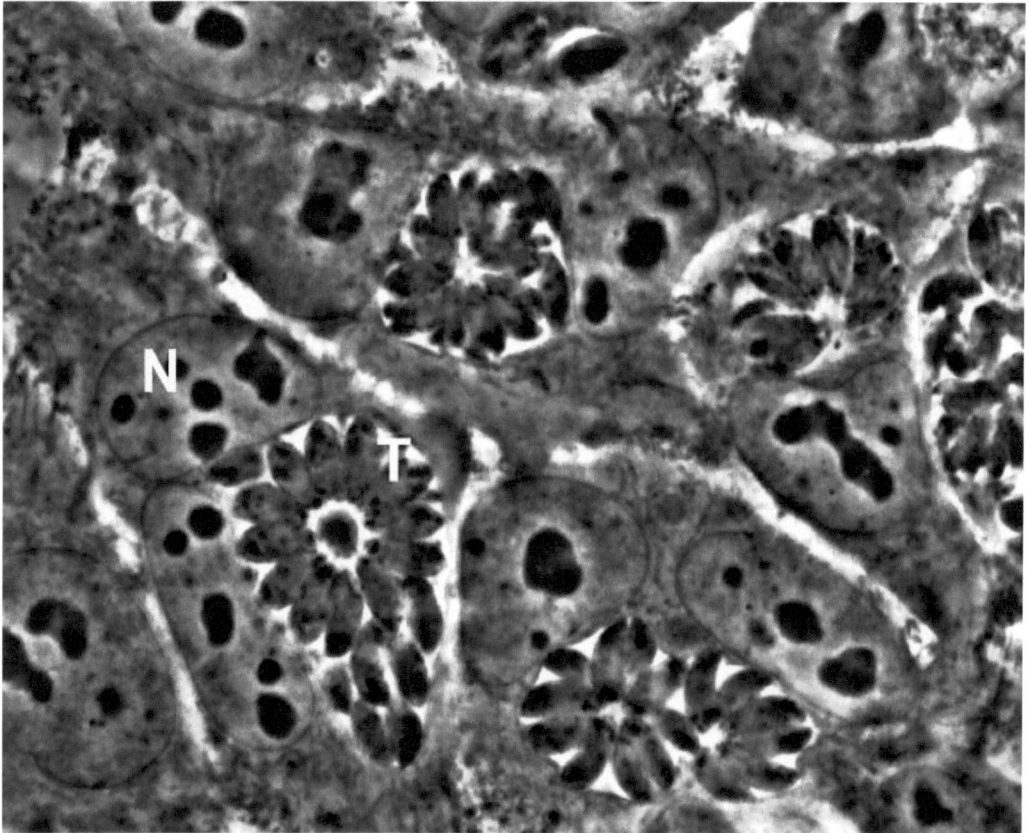

Fig. (1). Phase contrast image of *T. gondii* RH strain tachyzoites in Hep-2 cell culture forming rosettes inside the host cell. N Hep-2 cell nucleus, T tachyzoites. Image obtained at School of Medicine UNAM by Rivera-Fernández N.

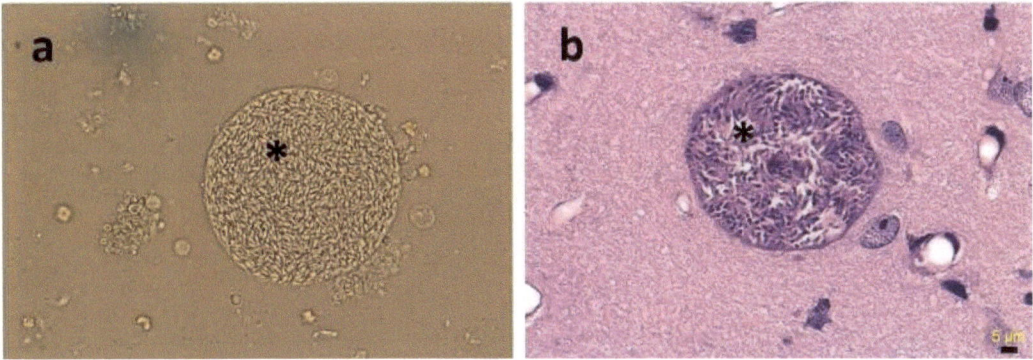

Fig. (2). *T. gondii* tissue cyst containing bradyzoites obtained from mice infected with *T. gondii* ME49 strain. a) Tissue cyst obtained from the brain suspension of an infected mouse. b) Mouse brain sagittal section stained with hematoxylin-eosin with a central tissue cyst. Asterix (tissue cyst). Image obtained at School of Medicine UNAM by Rivera-Fernández N.

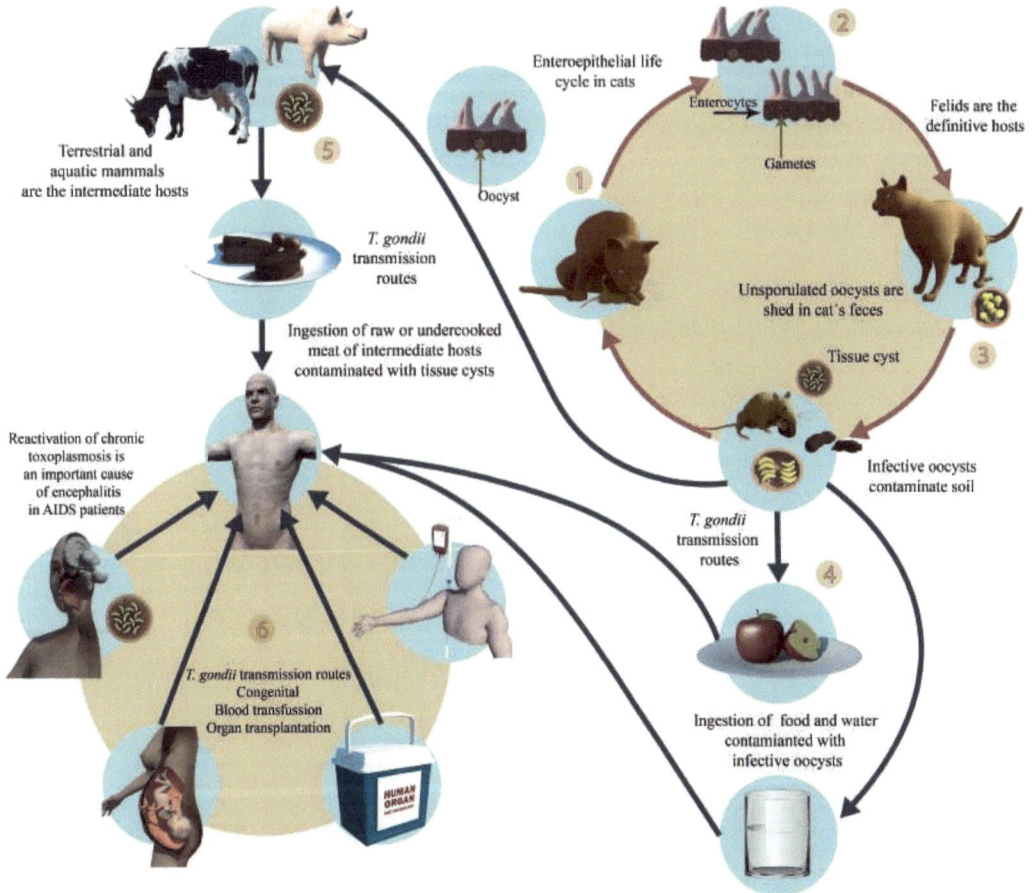

Fig. (3). *T. gondii* life cycle. Terrestrial and aquatic mammals including humans, as well as birds, are *T. gondii* intermediate hosts. Small mammals infected with *T. gondii* tissue cysts containing bradyzoites (located mostly in muscle tissue) are eaten by a domestic cat (definitive host) (1). Bradyzoites invade small intestine enterocytes and proliferate in order to develop gametes (2), which give rise to oocysts that are released into the environment with the cat's feces (3), and under suitable temperature and humidity conditions, they become infective about five days after being released, contaminating food and water (4). When oocysts are ingested by an intermediate immunocompetent host, *T. gondii* acute infection initiates and eventually parasites develop into tissue cysts starting a chronic infection (5). Humans can be infected by consuming food or water contaminated with oocysts or by raw or uncooked meat infected with tissue cyst. Other less frequent mechanisms of infection are blood transfusions, organ transplantation or vertical transmission. *T. gondii* can cause abortion in pregnant women and encephalitis in immunocompromised patients, among other severe medical complications (6). Figure designed by Rivera-Fernández N.

Immunocompetent hosts with primary toxoplasmosis do not require treatment unless a visceral compromise is present. The treatment of choice to control toxoplasmosis is a combination of pyrimethamine and sulfadiazine, which inhibits parasite folate metabolism, and to treat congenital toxoplasmosis spiramycin can be used [19]. These drugs decrease the replication of *T. gondii* tachyzoites and can induce interconversion; nevertheless, they are active only during the acute phase of the infection and can cause severe side effects such as hepatic necrosis and thrombocytopenia in patients. In addition to this, resistance has also been reported [20]. There are no pharmacological alternatives to treat chronic infections caused by tissue cysts and no human vaccines are available to control toxoplasmosis yet [20]. Therefore, high efficiency and low-toxicity drugs against *T. gondii* are required and natural compounds could provide a viable alternative to achieve this goal.

Table 1. Toxoplasmosis oral treatment options [8, 13, 16].

Acute toxoplasmosis	Pyrimethamine (25 mg/day), sulfadiazine (100 mg/day), folinic acid (5-15 mg/day) for 3-4 weeks.
Ocular toxoplasmosis	Pyrimethamine (25-50 mg/day), sulfadiazine (550-1000 mg/day), folinic acid (5-15 mg/day). 4-6 weeks.
Pregnant women	1st trimester spiramycin 1g every 8 h (it can be given throughout the pregnancy) Up to 18th weeks pyrimethamine 50-100 mg/day every 12 h; sulfadiazine 75 mg/kg/day; folinic acid 15 mg/day (it can be given throughout the pregnancy).
Newborn treatment	Pyrimethamine (2 mg/Kg/day for 2 days, then 1 mg/Kg/day for 6 months, then same dose 3 times per week), sulfadiazine (0.5 mg/Kg twice daily), folinic acid (25mg/ twice per week). For 6 months to one year.
Cerebral toxoplasmosis	Pyrimethamine 50-74 mg/day, sulfadiazine 1000-1500 every 6 h, folinic acid 10-20 mg/day.
Alternative drugs for toxoplasmosis treatment: Clindamycin, atovaquone, trimethoprim/ sulfamethoxazole.	

ETHNOBOTANICAL RECORDS

Medicinal plants are an important source of biologically active compounds, many of which can be considered as models for the synthesis of drugs useful in the treatment of various diseases [3]. According to the World Health Organization (WHO, 2020), 2/3 of the people residing in countries with peripheral economies use plants as a complementary system to allopathic medicine. Within traditional medicine, a great variety of plants with potential antiparasitic activity have been described, without accurate records of the causative parasites. Toxoplasmosis is not a disease commonly identified in traditional medicine; nevertheless, some records relate to the use of medicinal plants for its treatment. On the island of Madagascar, the Mahabo-Mananivo community use leaves, fruits, seeds, and roots obtained from *Carica papaya* L. (known as "paza") to treat *T. gondii* infection [21].

In vitro Activity Against *T. gondii*

Looking for new treatments to control *T. gondii* infections, some research groups have focused on the evaluation of *in vitro* toxoplasmicidal effects of several medicinal plants.

Pleopeltis Crassinervata

Hexane fractions from *Pleopeltis crassinervata* fronds were evaluated against *T. gondii* RH strain tachyzoites which exhibited an IC_{50} of 16.90 mg/mL; cytotoxicity of hexane fractions was also evaluated in Hep-2 cells and neuroblasts resulting in non-cytotoxicity in both cell lines. Although no isolated compounds were reported, toxoplasmic activity is attributed to terpenoid-like compounds [22].

Vernonia Colorata

The *in vitro* activity of nine plants used in traditional West African medicine against RH *T. gondii* tachyzoites was described in 2000. *Azadirachta indica*, *Cinnamonnum camphora*, *Lippia multiflora*, *Vernonia colorata*, *Guiera senegalensis*, *Combretum micranthum*, *Ximenia americana*, *Cochlospermum planchonii* and *Sida acuta* were collected in Ivory Coast and aqueous extracts (infusion and/or decoction) were prepared from different plant organs. When evaluating the *in vitro* effects on the growth of *T. gondii* in MRC5 fibroblast tissue cultures, it was observed that the plant with the highest activity was *V. colorata* with an IC_{50} of 17 and 18 µg/mL for infusion and decoction (obtained from stems and leaves) respectively, with low toxicity to the host cell [23].

Sophora Flavescens and Torilis Japonica

Hydroalcoholic extracts prepared from whole plants (root, stem and leaf) of five species: *Sophora flavescens*, *Sinomenium acutum*, *Pulsatilla koreana*, *Ulmus macrocarpa,* and *Torilis japonica* were evaluated *in vitro* against RH *T. gondii* tachyzoites grown on equine dermal cells at concentrations of 0.0195 to 0.625 μg/mL by the 3H-uracil method. The extracts obtained from *Sinomenium Acutum*, *Pulsatilla koreana,* and *Ulmus macrocarpa* were toxic at the highest concentration and their inhibitory effect was less than 66%. *T. japonica* inhibited the proliferation of *T. gondii* between 99.3% and 53% at concentrations between 0.156 and 0.0195 μg/mL, respectively. *S. flavescens* extract inhibited the growth of *T. gondii* by 98.7% and 27.2% at the same concentration range as *T. japonica*. The inhibitory activity of the extracts was higher than that obtained with 0.050 μg/ml of pyrimethamine [24].

The fractions obtained from the alcoholic extracts of *T. japonica* and *S. flavescens* were evaluated against RH tachyzoites proliferation in equine dermal cells at concentrations between 2,850 and 0.356 ng/mL. The fraction with the highest anti-*Toxoplasma* activity (99%) was SF1 obtained from *S. flavences*, whose activity was higher than that obtained with sulfadiazine at the same concentrations (97%). In this study, pure compounds were not isolated [25].

Ginkgo Biloba

The activity of ginkgolic acids (GAS) extracted from the sarcotesta of *Ginkgo biloba* seeds, as well as compounds isolated by column chromatography, was evaluated against purified RH *T. gondii* tachyzoites using azithromycin as a reference drug. At a concentration of 100 μg/mL, GAS presented a significant toxoplasmicidal activity close to 100%, and inhibition of protein and DNA synthesis was observed [26]. GAS and azithromycin did not show cytotoxic effects on human fibroblasts at concentrations below 100 μg/mL.

Azadirachta Indica and Melia Azedarach

Azadirachta indica and *Melia azedarach,* commonly known as neem and cinnamon respectively, are species belonging to the Meliaceae family. The effect of the leaves aqueous extracts was evaluated against *T. gondii* RH tachyzoites (maintained in Vero cell culture) at concentrations from 0.15 to 5 mg/mL. Aqueous neem extract at 5 mg/mL reduced the parasite invasion rate to 85%. Cinnamon extract, at 0.5 and 1 mg/mL, reduced parasite invasion by 70% and 90%, respectively. All concentrations were non-toxic in Vero cells. Azadirachtin as well as some other compounds detected in the aqueous extracts of neem and cinnamon are limonoid type, so it would be of interest to carry out structure-activity studies with these molecules in order to know the probable mechanism of action that affects parasite invasion and proliferation. The ultrastructural changes of the tachyzoites exposed to these extracts were evaluated by transmission electron microscopy. A drastic disorganization of the secretory system was observed. This lesion could be due to the effect of some of the chemical components in the tachyzoites membranes that are high in lipids [27].

Psidium Guajava and Tinospora Crispa

The activity of *Tinospora crispa* stem bark methanolic extract and *Psidium guajava* leaves aqueous extract was evaluated against *T. gondii* RH tachyzoites maintained in Vero cells, at 3-200 µg/mL using clindamycin as a reference drug. In this study, IC_{50} values of 7.7 µg/mL for *T. crispa* and 121 µg/mL for *P. guajava* were obtained compared to those of the clindamycin-treated group that presented an IC_{50} of 6.24 µg/mL. *T. crispa* showed the best toxoplasmicidal activity, then the extract was fractionated by column chromatography to evaluate its activity. Of the 8 fractions obtained, F5 was the most active, even better than clindamycin, presenting an IC_{50} of 6.06 µg/mL. Toxoplasmicide activity was attributed to the presence of alkaloids found in the methanolic extract [28].

Eurycoma Longifolia

Two fractions obtained from *Eurycoma longifolia* methanolic extract (TAF 355 and TAF 401) were evaluated on RH tachyzoites maintained in Vero cell culture. IC_{50} values of 0.016 µg/mL and 0.369 respectively were obtained, in comparison to 0.883 µg/mL for reference drug (clindamycin). Fractions decreased the confluence of the host cells at 36 hours post-exposure [29]. *In situ* studies demonstrated that the active fractions of *E. longifolia* showed potent antiproliferative activity on tachyzoites with an IC_{50} of 1,125 µg/mL for TAF 355, and 1,375 µg/mL for TAF 401 fraction [30].

Myristica Fragrans

The effect of *Myristica fragrans* seed essential oil (walnut mozcada) was evaluated on RH tachyzoites maintained in Vero cells. The culture was treated with different concentrations of the essential oil and 24 hours after treatment its efficacy was evaluated by the tetrazolium salt (3-(4,5-dimethylthiazol-2-yl-
-2,5-diphenyltetrazolium bromide (MTT) assay obtaining an EC50 of 24.45 µg/mL, in comparison with the EC50 obtained with the reference drug clindamycin (16.57 µg/mL). It was concluded that *M. fragrans* essential oil has a strong anti-*T. gondii* activity and causes low cytotoxic effects in the host cells [31].

Astragulus Membraneceus and Scutellaria Baicalensis

Yang *et al.* [32] conducted studies concerning the efficacy of *Astragulus membraneceus* and *Scutellaria baicalensis* roots aqueous extracts on RH tachyzoites expressing the green fluorescent protein (GFP) in HeLa cells (human cervical carcinoma). These plants are used in traditional Chinese medicine as immunostimulatory and anti-inflammatory agents [32]. Tachyzoites morphology changes were also evaluated at different incubation times (72, 96, and 120 h). Extracts were used at 1 to 10 mg/mL concentrations and trimethoprim/sulfamethoxazole was used as a reference drug. Both extracts reduced intracellular replication of *T. gondii* in less percentage when compared with that of the reference drug.

Sambucus Nigra

Sambucus nigra is used in Iran for its repellent, antibacterial, antiprotozoal, and anti-hemorrhoid properties. Methanolic extracts prepared with fruits and leaves of this plant proved to be effective against purified extracellular *T. gondii* RH tachyzoites. Results showed that the methanolic fruit extract at 5 and 10 mg/mL had a 100% toxoplasmicide efficacy after incubation for 60 and 120 minutes, respectively. At higher concentrations (25 and 50 mg/mL), the same effect was observed 30 minutes after treatment. A mortality of 98% was observed after 180 minutes of treatment with the leaf extract (100 mg/mL) [33].

Balsamociteus Camerunensis

Some compounds obtained by column chromatography were isolated from the dichloromethane/methanolic extract fractions of *Balsamociteus camerunensis*: marmina (HEN1), xanthoxyletin (HEN3), 6,7-dimethoxycoumarin (HEN4) and 1-hydroxy-3-methoxy-acridone (HEN5). These compounds inhibited the growth of *T. gondii* RH tachyzoites in 5, 82.12, 46.44, and 22.03% for HEN1, HEN3, HEN4 and HEN5 respectively, in human foreskin fibroblasts (HFF) cells. HEN3 was the compound with the highest activity [34].

Jatropha Curcas

Jatropha curcas (Jc) is used in Brazil based on its diuretic and anti-leukemic properties. The effect of Jc seed extract (JcSE) was evaluated against RH tachyzoites at different concentrations (0.01 to 5.0 µg/mL). Results showed that JcSE reduced parasite invasion in Vero cells to 26.2% at 3.0 µg/mL and the number of intracellular parasites to 18%. The extract was intraperitoneally injected to mice at 90 mg/kg and no toxic signs were observed. Fractionation was performed, and a cysteine protease inhibitor was found, which could be responsible for the extracellular proteolysis of the parasites. There are five enzymes of the C1 cysteine proteases family that participate in the growth and survival of *T. gondii*; therefore, this inhibitor can cause morphological and parasitophorous vacuole changes. No reference drugs were included in this study [35].

Shorgum Bicolor

Abugri *et al.* [36] evaluated S*horgum bicolor* extracts obtained by maceration, fractions obtained by column chromatography, and two identified compounds in RH tachyzoites maintained in HFF cells. The most active extract was ethyl acetate with an IC_{50} of 1.39 µg/mL. Fractions showed IC_{50} values between 0.36 to 7 µg/mL. Luteonidine chloride was the most active isolated compound with an IC_{50} of 0.66 µg/mL [36].

Achillea Millefolium and hypericum Perforatumen

Ethanolic extracts from *Achillea millefolium* aerial parts and *Hypericum perforatumen* were evaluated on HeLa cells infected with RH tachyzoites. Both extracts showed toxoplasmicidal effects reaching up to 100% mortality at 100 µg/mL concentration. However, at low concentrations (10 µg/mL) the mortality rate did not exceed 6%, being of 5.31% and 5.11% for *H. perforatum* and *A. millefolium,* respectively. Evaluation in cell line showed EC_{50} of 215 and 153 µg/mL, respectively in comparison with that of pyrimethamine (EC_{50} of 0.176 µg/mL). *H. perforatum* showed a better anti-*Toxoplasma* activity than that of *A. millefolium*; however, it was concluded that these extracts do not possess a significant anti-*Toxoplasma* activity [37]. In other studies, it was concluded that ethanolic extracts of *Myrtus communis* and *Artemisia auchery* are not toxoplasmicidal candidates [38].

Cola Gigantea, Tectona Grandis and Vernonia Amigdalina

Tectona grandis and *Vernonia amigdalina* are medicinal plants from Africa. Hydroalcoholic and ethanolic extracts from the leaves and bark of both plants were evaluated *in vitro* against *T. gondii*. Using a colorimetric method with RH---gal strain encoding galactosidase, parasite inhibition in HFF cell culture was determined after 96 hours of exposure. The most active extract was *V. amigdalina* leaf ethanolic with an IC_{50} of 5.6 µg/mL [39]. With a similar method, the activity of the seeds essential oil of *Cola gigantea* was evaluated in *T. gondii* RH-F2 strain at very high concentrations. After 72 hours of incubation in 200 µg/mL, the viability percentage was close to zero as well as the viability of uninfected cells exposed to the same IC_{50} concentrations [40].

Anti-*T. gondii in vivo* and *in vitro* activity

Zingiber Officinale (ginger), Ginkgo Biloba and Glycyrrhiza Glabra

In Korea, the effects of *Zingiber officinale* (ginger), *Ginkgo biloba* and *Glycyrrhiza glabra* methanolic extracts, and the GE/F1 fraction obtained from ginger methanolic extract were evaluated on cell proliferation and RH tachyzoites viability using the MTT assay in C6 cells (mouse glioma). The concentration range was from 30 to 240 μg/mL. Additionally, the effect of the fraction on the survival and production of cytokines such as interferon-gamma (IFN- ɣ) and interleukin 8 (IL-8) was determined in BALB/c mice. Ginger extract (GE) inhibited the proliferation and viability of C6 cells infected with *T. gondii* at concentrations between 60 and 120 μg/mL, while the GE/F1 fraction showed a strong antiproliferative activity at 240 μg/mL. The toxoplasmicidal efficacy was greater than the one obtained with sulfadiazine at the same concentrations. Results showed that GE/F1 fraction presented activity on the parasitophorous vacuole and blocked parasite-induced C6 cell apoptosis; this activity is attributed to the decrease in the expression of P21, P53, caspase-3, and bax. The GF/F1 fraction (500 μg/mL) inhibited the proliferation and growth of *T. gondii*, increased the survival of the infected mice, and decreased the concentrations of IFN-ɣ and IL-8. No reference drugs were included in these studies [41].

Thymus Broussonetii

Thymus broussonetii essential oils obtained by hydrodistillation were evaluated against *T. gondii*. This is an endemic medicinal plant from Morocco mainly used for its antidiarrheal, antipyretic, and toning functions. Experimentally, it has been observed that *T. broussonetii* has antibacterial and antifungal properties. The effect of its essential oil was evaluated in OF1 female mice strain orally infected with tissue cysts of *T. gondii* Prugniaud strain and treated with a single dose of 20 μg. The results showed a reduction in tissue cysts in the brain. The mechanism of action could be due to the lysis of the bradyzoites released in the intestine by the action of the essential oil components [42].

Bunium Persicum

The prophylactic and therapeutic effects of *Bunium persicum* essential oil were evaluated in a murine model with acute toxoplasmosis. This oil was obtained by hydrodistillation from plant seeds. NMRI strain male mice were used for this assay. To evaluate the prophylactic effect, mice were treated with the essential oil at concentrations of 0.05 and 0.1 mL/kg for 14 days. Twenty-four hours post-treatment, the animals were intraperitoneally infected with RH tachyzoites. To assess the therapeutic effect, mice were infected in the same way as the animals in the prophylactic study and were treated every 12 hours for five days with 0.05 and 0.1 mL/kg of the essential oil, dispensing the first dose 24 hours post-infection. Mice treated with 0.05 and 1 mL/kg of the essential oil showed survival of four to five days respectively, in comparison to that of the control group in which the animals died on day 5 after infection. Some oil components are γ-terpineno (46.%), cuminaldehyde (15.5%), ρ-cimeno (6.7%) and limonene (5.9%) [43].

Thymus Vulgaris

In Egypt, the efficacy of the ethanolic extract obtained from *Thymus vulgaris* leaves (TVE), a medicinal plant of wide distribution used in the Middle East for its antispasmodic, anti-cough, anti-broncholytic, analgesic, diuretics, anti-inflammatory, antihelmintic and sedative properties was evaluated against *T. gondii*. It is known that its extracts have antiparasitic activity against other parasites, including *Trypanosoma cruzi* [44]. The therapeutic and prophylactic effects of TVE extract were evaluated in Swiss albino mice with chronic toxoplasmosis at 500 mg/kg, using sulfadiazine-pyrimethamine (SP) at doses of 200 and 12.5 mg/kg/day, respectively, as reference drug. In the control group, 630 tissue cysts were obtained. In the group prophylactically treated with the extract, the number of cysts decreased by 24% while the reduction in the group therapeutically treated was 46% compared to a 51% reduction in the number of cysts obtained in the SP group. Animals treated prophylactically with the extract presented mild histopathological changes in the brain. Unlike the therapeutically treated mice, whose number of tissue cysts decreased, as well as did the histological changes [44].

Echinaceae Purpurea

The effect of *Echinaceae purpurea* aerial parts and flowers infusion (AEEP) was evaluated at doses of 30, 100, and 300 mg/kg in Swiss mice infected with *T. gondii* RH strain. Toxoplasmicidal efficacy was obtained by counting parasites purified from intraperitoneal and liver lavages with PBS. All animals treated with 300 mg/kg of AEEP showed a reduction in the RH tachyzoites number obtained by peritoneal lavage (29 tachyzoites) compared with that of the control group without treatment (300 tachyzoites). In the treated animals, the number of liver tachyzoites significantly decreased when compared to that of the non-treated control group. These results indicate that the extract has some protective function against parasite proliferation [45].

Nigella Sativa

It has been reported that *Nigella sativa* oil has a potent immunostimulatory effect, which increases gamma interferon levels and potentiates T cells against infectious diseases and cancer. This oil was evaluated against *T. gondii* RH strain tachyzoites in Swiss male mice with three different protocols, 12.5 mg/kg of essential oil, 12.5 mg/kg of essential oil plus 5 mg/kg of pyrimethamine and clindamycin/pyrimethamine at 25 and 12 mg/ kg/day respectively. All animals were treated every 24 hours for five days. Animals treated with oil plus pyrimethamine and clindamycin-pyrimethamine survived beyond day 10 post-infection (PI). Only 30% of the animals treated with the oil, survived to the eight-day PI, compared to infected animals without treatment, which died at day five PI. The number of tachyzoites obtained after treatment in the control group was 14 in the liver and eight in the spleen, compared with the number of parasites from the group treated with clindamycin-pyrimethamine, which showed a significant reduction with only 2.3 and 2.6 liver and spleen tachyzoites, respectively. Fourteen tachyzoites were found in the liver and seven in the spleen of the animals treated with oil, in comparison with those of the group treated with oil/pyrimethamine, where 2.9 and 2.6 tachyzoites were obtained from liver and spleen, respectively. These results showed that the oil alone has not a toxoplasmicidal effect, while the combination with pyrimethamine enhances its efficacy, which is comparable with that obtained by the combination of clindamycin/pyrimethamine [46].

Berberis Vulgaris and Saturja Khuzestanica

The prophylactic effect of *Berberis vulgaris* root methanolic extract (1 and 2 g/kg/day) and *Saturja khuzestanica* essential oil (0.2 and 0.3 mL/ kg/day) was evaluated for fourteen days in NMRI mice infected with *T. gondii* RH strain. The parasitic load was counted by peritoneal lavage, the control group showed 288 x 10^4 tachyzoites, while with the methanolic extract 131 $\times$ 10^4 and 79 $\times$ 10^4 tachyzoites were observed with both doses. The survival rate of the animals treated with *B. bulgaris* was of nine days when compared with that of the control group (14 days). 260 x 10^4 and 189 x 10^4 tachyzoites were observed with the doses used to evaluate the essential oil; the survival rate lasted 7 days with respect to the untreated control in which mice died on the 5th day PI [47, 48].

Artemisia Annua

Artemisia annua is an endemic plant in northern China that has been used to develop novel antimalarials. Based on this background, the effect of an infusion of *A. annua* aerial parts was tested on RH *T. gondii* tachyzoites invasive properties in HFF cells and on C57BL/6 female mice infected with RH and ME49 strains. Sulfadiazine was used as a reference drug. In both cases, the IC_{50} was determined. Artemisinin infusion concentration was 0.2%, quantified by high-performance liquid chromatography (HPLC). In all cases, HFF cell viability was greater than 72%, even when the highest concentrations of both treatments were tested (10,000 and 200 µg/mL, respectively). *In vitro* assays demonstrated that *A. annua* infusion and sulfadiazine inhibited the growth of *T. gondii*, with IC_{50} values of 95 µg/mL and 3 µg/mL, respectively. *In vivo* tests indicated that *A. annua* infusion and sulfadiazine treatment effectively controlled the infection with ME49 strain (100% survival in treated animals, 30 days after infection) but could not control the infection with the virulent RH strain (20-50% survival). In this study, the effect of the infusion on the production of Interleukin 12 (IL-12), Tumor Necrosis Factor (TNF), and nitric oxide (NO) in peritoneal macrophages was also evaluated. The results showed that the *A. annua* infusion has a stimulating effect on the production of IL-12, a cytokine related to the control of the parasite. *A. annua* infusion demonstrated good efficacy in the control of infection with *T. gondii* [49].

Piper Betle, P. Sarmentosum and P. Nigrum

In recent studies, the effect of ethanolic extracts from *Piper betle* and *P. sarmentosum* leaves and *P. nigrum* seeds was evaluated as well as toxicity in HFF cell culture at concentrations from 1 to 100 µg/mL. Sulfadiazine was used as a reference drug at 0.01 to 1 µg/mL. The effect of the extracts on RH intracellular tachyzoites was evaluated in the same cell line at 1 to 100 µg/mL and in extracellular tachyzoites at 25 µg/mL. For the *in vivo* tests, C57BL female mice infected with 1×10^3 RH tachyzoites were used. Twenty-four hours post-infection, these animals were treated with the extracts at 25 to 400mg/kg/day. The most active extract against *in vitro T. gondii* was *P. betle* with an IC_{50} of 32.3 µg/mL. The IC_{50} of the other extracts was greater than 100 µg/mL. All extracts showed an effect on intracellular tachyzoites at concentrations higher than 25 µg/mL in comparison with the lowest concentration of sulfadiazine (1.0 µg/mL) required to affect the parasite. In the *in vivo* trial, animals treated with 400 and 100 µg/kg of *P. betle* extract reduced their clinical signs with a survival percentage of 100% and 83%, respectively. There were no significant differences between the number of parasites obtained from the brain of the surviving mice treated with *P. betle* or sulfadiazine. It was concluded that the *P. betle* ethanolic extract exhibits toxoplasmicidal activity in *in vitro* and *in vivo* models and increases the survival rate of mice treated with 400 µg/kg [50].

Allium Paradoxum, Feijoa Sellowiana and Quercus Castaneifolia

Methanolic extracts of the leaves and fruits of *Allium paradoxum, Feijoa sellowiana* and *Quercus Castaneifolia* were evaluated *in vitro* and *in vivo* against RH *T. gondii* strain by the MTT assay. Vero cells were infected with the parasite and exposed to concentrations up to 400 µg/mL for 24 hrs. The best activity was observed with *F. sellowuiana* leaves methanolic extract with an IC_{50} of 12.77 μg/mL that also improved the survival of BALB/c mice treated with 200 mg/kg/day for 11 days PI in comparison with the non-treated animals which died 8 days after infection [50].

Araucaria Heteropil

The main component of the *Araucaria heteropil* (AHR), 13-epi-cupresic (CUP) resin has activity against *T. gondi* RH extracellular tachyzoites (parasites harvested from intraperitoneal macrophages). Viability of tachyzoites was determined by the trypan blue dye exclusion method with an EC_{50} of 3.69 μg/mL. When tested in chronic and acute toxoplasmosis models, animals treated with CUP showed an increase in the survival rate of approximately 70% at 370 mg/kg [51].

Aloe Vera and Eucaliptus

Aloe vera and *Eucaliptus* methanolic extracts have been tested by the MTT method in *T. gondii* RH. IC_{50} values of 13.2 and 24.7 μg/mL were obtained, respectively. Survival percentage was determined in an acute toxoplasmosis model, *Aloe vera* extract (50 mg/kg/day) increased the survival rate by 60% at 9 days PI with respect to the untreated group that died at day 7 post-infection PI [52].

Extracts, fractions, and isolated compounds evaluations are summarized in Tables **2 - 5.**

Table 2. Plants with activity against *T. gondii*.

Species / Extract	Family	Used Part	*T. gondii* Inhibition CI_{50} [μg/mL]	Citotoxicity C_{50} [μg/mL]	Refs.
Achillea millefolium -Ethanolic	Asteraceae	Stem Leaf	-	-	[19]
Allium paradoxum -Methanolic	Amaryllidaceae	Leaf	212.2	148.5	[35]
Aloe vera -Methanolic	Asphodelaceae	Leaf	13.2	43.2	[35]
Araucaria heterophylla	Araucariaceae	Resin	3.9	-	[34]
Artemisia annua -Infusion	Asteraceae	Stem Leaf	95	-	[32]

(Table 2) cont.....

Species / Extract	Family	Used Part	T. gondii Inhibition CI_{50} [μg/mL]	Citotoxicity C_{50} [μg/mL]	Refs.
Artemisia aucheri -Ethanolic	Asteraceae	Aerial parts	-	-	[21]
Astragalus Membranaceus -Aqueous	Fabaceae	Root	-	-	[14]
Azadirachta indica -Infusion -Decoction	Meliaceae	Stem Leaf	>1000 494	>1000 >1000	[5]
Berberis vulgaris -Methanolic	Berberidaceae	Root	-	-	[30]
Bunium persicum	Apiaceae	Seed	-	-	[26]
Cinnamonnum camphora -Infusion -Decoction	Lauraceae	Bark	789 565	1000 1000	[5]
Cochlospermum plantonii -Infusion	Bixaceae	Tuber	>1000	1000	[5]
Cola gigantea -Essencial oil	Sterculiaceae	Seed	-	-	[23]
Combretum micrantbum -Infusion -Decoction	Combretaceae	Stem Leaf	217 254	>1000 >1000	[5]
Echinacea purpurea -Infusion	Asteraceae	Flower Stem Leaf	-	-	[28]
Eucaliptus -Methanolic	Myrtaceae	Leaf	23.7	58.7	[35]
Feijoa sellowiana	Myrtaceae	Leaf Fruit	12.77 180.2	77.36 236.5	[35]
Ginkgo biloba	Ginkgoaceae	Leaf	-	-	[24]
Glycyrrhiza glabra	Fabaceae	Root	-	-	[24]
Guiera Senegalensis -Infusion -Decoction	Combretaceae	Leaf	351 177	500 250	[5]
Hypericum perforatum -Ethanolic	Hypericaceae	Stem Leaf	-	-	[19]
Jatropa curcas -Protein extract	Euphorbiaceae	Seed	-	-	[17]
Lippia multiflora -Infusion -Decoction	Verbenaceae	Leaf	201 127	>1000 >1000	[5]

(Table 2) cont.....

Species / Extract	Family	Used Part	T. gondii Inhibition CI_{50} [μg/mL]	Citotoxicity C_{50} [μg/mL]	Refs.
Myristica fragrans -Essential oil	Myristicaceae	Fruit	24.45	24.8	[13]
Myrtus communis -Ethanolic	Myrtaceae	Leaf	-	-	[20]
Nigella sativa -Essential oil	Ranunculaceae	-	-	-	[29]
Psidium guajava -Aqueous	Myrtaceae	Leaf	121	110	[10]
Piper betle, P. nigrum y P sarmentosum -Ethanolic	Piperacea	Leaf Seed	-	-	[33]
Quercus Castaneifolia -Methanolic	Fagaceae	Fruit	74.73	26.81	[33]
Sambucus nigra -methanolic	Caprifoliaceae	Fruit Leaf	-	-	[15]
Saturja khuzestanica -Essential oil	Lamiaceae	Aerial parts	-	-	[31]
Scutellaria baicalensis -aqueous	Fabaceae	Root	-	-	[14]
Sida acuta -Infusion	Malvaceae	Flower Leaf	>1000	>1000	[5]
Sorghum bicolor -Ethanolic	Poaceae	Leaf	20.38	46.67	[18]
Tectona grandis -Hydroalcoholic -Ethanolic -Hydroalcoholic -Ethanolic	Verbenaceae	Leaf Leaf Bark Bark	59.8 143.3 176.7 15.3	45.9 38.4 85.2 147.7	[23]
Thymus broussonetii -Essential oil	Lamiaceae	Stem Leaf	-	-	[25]
Thymus vulgaris -Ethanolic	Lamiaceae	Leaf	-	-	[27]
Tinospora crispa -Methanolic	Menispermaceae	Stem	7.7	147	[10]
Vermonia Colorata -Infusion -Decoction	Composeae	Stem Leaf	17 18	250 250	[5]
Vernonia amygdalina	Asteraceae	Leaf	5.6	303.6	[23]

(Table 2) cont.....

Species / Extract	Family	Used Part	*T. gondii* Inhibition CI_{50} [µg/mL]	Citotoxicity C_{50} [µg/mL]	Refs.
Ximenia americana -Infusion -Decoction	Oleaeae	Stem Leaf	>1000 469	1000 1000	[5]
Zingiber officinale	Zingiberaceae	Root	-	-	[24]
Reference drugs -Pyrimethamine -Chloroquine -Clindamycin -Clindamycin -Sulfadiazine	-	-	0.04 ND 6.24 - 3	50 ND 613 16.57 -	

Table 3. Growth inhibition percentage of some plants with *in vitro* anti-*Toxoplasma* activity.

Species / Extract	Family	Used Part	*T. gondii* Inhibition (%)	[µg/mL]	Ref
Azadirachta indica -aqueous	Meliaceae	Leaf	85	5000	[9]
Melia azedarach -aqueous	Meliaceae	Leaf	90	1000	[9]
Pulsatilla koreana -Hydroalcoholic	Ranunculaceae	Root, Stem, Leaf	99.7	0.312	[6]
Sinomenium acutum -Hydroalcoholic	Menispermaceae	Root, Stem, Leaf	63.4	0.312	[6]
Sophora flavescens -Hydroalcoholic	Fabaceae	Root, Stem, Leaf	3.4	0.312	[6]
Torilis japonica -Hydroalcoholic	Apiaceae	Root, Stem, Leaf	24.5	0.312	[6]
Ulmus macrocarpa -Hycroalcoholic	Ulmaceae	Root, Stem, Leaf	98.5	0.312	[6]
Reference drug -Sulfadiazine - Pyrimethamine	-	-	9.7 27.2	0.312 0.312	[6]

Table 4. Compounds isolated and evaluated *in vitro* against *T. gondii* and inhibition growth percentage.

Species	Family	Identified Compound	*T. gondii* Inhibition IC_{50} [μg/mL]	IC_{50} Cytotoxicity [μg/mL]	Ref
Balsamocitrus camerunensis	Rutaceae	- marmina, -xantoxiletine - 6, 7-dimethoxycoumarin -1-hydro-y-3-methoxy-acridone	- - - -	- - - -	[16]
Sorghum bicolor	Poaceae	-Luteolinidine -7-methoxyapigenidine	0.66 1.48	6.54 7.99	[18]

Table 5. Fractions evaluated *in vitro* against *T. gondii*.

Species	Extract	Fraction	*T. gondii* Inhibitión CI_{50} [μg/mL]	IC_{50} Citotoxicity [μg/mL]	Ref
Eurycoma longifolia	methanolic	TAF 355 TAF 401	0.369 0.882	- -	[11]
Gingko biloba	-	GAS	-	-	[8]
Sorghum bicolor	-	F1-F4(C/M 10:1) F2-5 (chloroform) F ethyl acetate	0.36 7 1.39	20.20 20.39 7.56	[18]
Tinospora crispa	methanolic	F5	6.06	156	[10]
Pleopeltis crassinervata	Methanolic	hexane	16.7	240	[22]
Reference drug -Clindamycin	-	-	0.016	-	[11]

Table 6. Fractions evaluated against *T. gondii* and growth inhibition percentage.

Species	Extract	Fraction	Inhibition *T. gondii* %	[μg/mL]	Ref
Gingko biloba	-	GAS	-	-	[8]
Sophora flavescens	Hydroalcoholic	SF1 SF2 SF3 SF4 SF5	99.6 96.9 88.5 94.9 82.1	0.00285 0.00285 0.00285 0.00285 0.00285	[7]
Torilis japonica	Hydroalcoholic	TJ1 TJ2	87.8 99.2	0.00285 0.00285	[7]
Reference drug -Sulfadiazine	-	-	97.0	0.00285	[7]

CONCLUDING REMARKS

Little research on medicinal plants as potential toxoplasmicides is done; so far, there are records of the evaluation of about 58 extracts, 14 fractions, and only seven compounds obtained from 53 species (belonging to 33 families) of plants used mainly in traditional medicine, evaluated in both *in vitro* and *in vivo* models. Six species of medicinal plants (*Tinospora crispa, Vermonia colorata, Sorghum bicolor, Pulsatilla koreana, Sinomenium acutum* and *Torilis japonica*) showed a relevant anti-*Toxoplasma* activity, with IC*50* <20 µg /mL *in vitro* models. *S. bicolor* compounds, Luteolinidine and 7-methoxyapigenidine, have proved to be the most active against *T. gondii* with IC_{50} values of 0.66 and 1.48 µg / mL, respectively. Until now, there are no active drugs against *T. gondii* tissue cysts, hence, there is an urgent need to find active molecules against this phase. According to the consulted data, *Echinaceae purpurea, Thymus vulgaris,* and *Thymus broussonetii* extracts were active in murine models chronically infected with *T. gondii* ME49 and Prugniaud strains; particularly, essential oil of *T. broussonetii* showed excellent results as a single dose of 20 µg prevented the development of tissue cysts in the host. Due to the high seroprevalence of *T. gondii,* the expanding number of immunocompromised patients, the adverse effects caused by current anti-*Toxoplasma* drugs like pyrimethamine/sulfadiazine combination, and the lack of human vaccines and active compounds against tissue cysts, the studies cited in this work encourage us to optimize natural products in order to find new anti-*Toxoplasma* lead compounds.

CONSENT FOR PUBLICATION

Not applicable.

CONFLICT OF INTEREST

The authors declare no conflict of interest, financial or otherwise.

ACKNOWLEDGEMENTS

The Authors thank Mrs Josefina Bolado, Head of Scientific Paper Translation Department, from Research Division, School of Medicine, UNAM for editing the English-language version of this manuscript.

REFERENCES

[1]   Rivera N, López PY, Rojas M, *et al.* Antimalarial efficacy, cytotoxicity, and genotoxicity of methanolic stem bark extract from *Hintonia latiflora* in a *Plasmodium yoelii* lethal murine malaria model. Parasitol Res 2014; 113(4): 1529-36.
      [http://dx.doi.org/10.1007/s00436-014-3797-9] [PMID: 24549754]

[2]   Newman DJ, Cragg GM. Natural products as sources of new drugs from 1981 to 2014. J Nat Prod

2016; 79(3): 629-61.
[http://dx.doi.org/10.1021/acs.jnatprod.5b01055] [PMID: 26852623]

[3]   Rios MY, Aguilar-Guadarrama AB, Navarro V. Two new benzofuranes from *Eupatorium aschenbornianum* and their antimicrobial activity. Planta Med 2003; 69(10): 967-70.
[http://dx.doi.org/10.1055/s-2003-45113] [PMID: 14648407]

[4]   Muñetón PP. Plantas medicinales: un complemento vital para la salud de los mexicanos. Entrevista con el Mtro. Erick Estrada Lugo Rev Dig Univ 2009; 10(9)

[5]   Mehlhorn H, Wu Z, Ye B. Treatment of Human Parasitosis in Traditional Chinese Medicine 2014.

[6]   Al Nasr I, Ahmed F, Pullishery F, El-Ashram S, Ramaiah VV. Toxoplasmosis and anti-*Toxoplasma* effects of medicinal plant extracts-A mini-review. Asian Pac J Trop Med 2016; 9(8): 730-4.
[http://dx.doi.org/10.1016/j.apjtm.2016.06.012] [PMID: 27569880]

[7]   Sharif M, Sarvi S, Pagheh AS, *et al.* The efficacy of herbal medicines against *Toxoplasma gondii* during the last 3 decades: a systematic review. Can J Physiol Pharmacol 2016; 94(12): 1237-48.
[http://dx.doi.org/10.1139/cjpp-2016-0039] [PMID: 27564395]

[8]   McFarland MM, Zach SJ, Wang X, *et al.* Review of experimental compounds demonstrating anti-*Toxoplasma* activity. Antimicrob Agents Chemother 2016; 60(12): 7017-34.
[PMID: 27600037]

[9]   Sepulveda-Arias JC, Veloza LA, Mantilla-Muriel LE. Anti-*Toxoplasma* activity of natural products: a review. Recent Pat Antiinfect Drug Discov 2014; 9(3): 186-94.
[http://dx.doi.org/10.2174/1574891X10666150410120321] [PMID: 25858302]

[10]   Rivera N, Mondragón FR. Cistogénesis de *Toxoplasma gondii*. Rev Ed Bioquím 2010; 29(1): 14-9.

[11]   Black MW, Boothroyd JC. Lytic cycle of *Toxoplasma gondii*. Microbiol Mol Biol Rev 2000; 64(3): 607-23.
[http://dx.doi.org/10.1128/MMBR.64.3.607-623.2000] [PMID: 10974128]

[12]   Munoz M, Liesenfeld O, Heimesaat MM. Immunology of *Toxoplasma gondii*. Immunol Rev 2011; 240(1): 269-85.
[http://dx.doi.org/10.1111/j.1600-065X.2010.00992.x] [PMID: 21349099]

[13]   Rivera Fernández N, Mondragón Castelán M, González Pozos S, *et al.* A new type of quinoxalinone derivatives affects viability, invasion, and intracellular growth of *Toxoplasma gondii* tachyzoites *in vitro*. Parasitol Res 2016; 115(5): 2081-96.
[http://dx.doi.org/10.1007/s00436-016-4953-1] [PMID: 26888289]

[14]   English ED, Striepen B. The cat is out of the bag: How parasites know their hosts. PLoS Biol 2019; 17(9)e3000446
[http://dx.doi.org/10.1371/journal.pbio.3000446] [PMID: 31487278]

[15]   Caballero-Ortega H, Uribe-Salas FJ, Conde-Glez CJ, *et al.* Seroprevalence and national distribution of human toxoplasmosis in Mexico: analysis of the 2000 and 2006 National Health Surveys. Trans R Soc Trop Med Hyg 2012; 106(11): 653-9.
[http://dx.doi.org/10.1016/j.trstmh.2012.08.004] [PMID: 22998951]

[16]   Wang JL, Elsheikha HM, Li TT, *et al.* Efficacy of antiretroviral compounds against *Toxoplasma gondii in vitro*. Int J Antimicrob Agents 2019; 54(6): 814-9.
[http://dx.doi.org/10.1016/j.ijantimicag.2019.08.023] [PMID: 31479744]

[17]   Hernández-Cortazar I, Acosta-Viana KY, Ortega-Pacheco A, Guzman-Marin EdelS, Aguilar-Caballero AJ, Jiménez-Coello M. Toxoplasmosis in Mexico: epidemiological situation in humans and animals. Rev Inst Med Trop São Paulo 2015; 57(2): 93-103.
[http://dx.doi.org/10.1590/S0036-46652015000200001] [PMID: 25923887]

[18]   Toledo González Y, Soto García M, Chiang Rodríguez C, Rúa Martínez R, Estévez Miranda Y, Santana Alas ER. Toxoplasmosis ocular. Rev Cuba Oftalmol 2010; 23: 812-26.

[19] Recursos en parasitología, Toxoplasmosis: UNAM 2010.
      http://microypara.facmed.unam.mx/?page_id=1573

[20] Montazeri M, Mehrzadi S, Sharif M, *et al.* Drug Resistance in *Toxoplasma gondii.* Front Microbiol
      2018; 9: 2587.
      [http://dx.doi.org/10.3389/fmicb.2018.02587] [PMID: 30420849]

[21] Razafindraibe M, Kuhlman AR, Rabarison H, *et al.* Medicinal plants used by women from Agnalazaha
      littoral forest (Southeastern Madagascar). J Ethnobiol Ethnomed 2013; 9: 73.
      [http://dx.doi.org/10.1186/1746-4269-9-73] [PMID: 24188563]

[22] Anacleto-Santos J, López-Camacho P, Mondragón-Flores R, *et al.* Anti-*Toxoplasma*, antioxidant and
      cytotoxic activities of *Pleopeltis crassinervata* (Fée) T. Moore hexane fraction. Saudi J Biol Sci 2020;
      27(3): 812-9.
      [http://dx.doi.org/10.1016/j.sjbs.2019.12.032] [PMID: 32127756]

[23] Benoit-Vical F, Santillana-Hayat M, Kone-Bamba D, Mallie M, Derouin F. Anti-*Toxoplasma* activity
      of vegetal extracts used in West African traditional medicine. Parasite 2000; 7(1): 3-7.
      [http://dx.doi.org/10.1051/parasite/2000071003] [PMID: 10743641]

[24] Youn HJ, Lakritz J, Kim DY, Rottinghaus GE, Marsh AE. Anti-protozoal efficacy of medicinal herb
      extracts against *Toxoplasma gondii* and *Neospora caninum.* Vet Parasitol 2003; 116(1): 7-14.
      [http://dx.doi.org/10.1016/S0304-4017(03)00154-7] [PMID: 14519322]

[25] Youn HJ, Lakritz J, Rottinghaus GE, *et al.* Anti-protozoal efficacy of high performance liquid
      chromatography fractions of *Torilis japonica* and *Sophora flavescens* extracts on *Neospora caninum*
      and *Toxoplasma gondii.* Vet Parasitol 2004; 125(3-4): 409-14.
      [http://dx.doi.org/10.1016/j.vetpar.2004.08.002] [PMID: 15482896]

[26] Chen SX, Wu L, Jiang XG, Feng YY, Cao JP. Anti-*Toxoplasma gondii* activity of GAS *in vitro.* J
      Ethnopharmacol 2008; 118(3): 503-7.
      [http://dx.doi.org/10.1016/j.jep.2008.05.023] [PMID: 18602775]

[27] Melo EJT, Vilela KJ, Carvalho CS. Effects of aqueous leaf extracts of *Azadirachta indica* A. Juss.
      (neem) and *Melia azedarach* L. (Santa Barbara or cinnamon) on the intracellular development of
      Toxoplasma gondii. Rev Bras Plantas Med 2011; 13: 215-22.
      [http://dx.doi.org/10.1590/S1516-05722011000200014]

[28] Lee W, Mahmud R, Noordin R, Piaru S, Perumal S, Ismail S. Alkaloids content, cytotoxicity and
      anti-*Toxoplasma gondii* activity of *Psidium guajava* L. and *Tinospora crispa.* Bang J Pharmacol 2012;
      7(4): 272-6.
      [http://dx.doi.org/10.3329/bjp.v7i4.12499]

[29] Kavitha N, Noordin R, Chan KL, Sasidharan S. *In vitro* anti-*Toxoplasma gondii* activity of root
      extract/fractions of *Eurycoma longifolia* Jack. BMC Complement Altern Med 2012; 12: 91.
      [http://dx.doi.org/10.1186/1472-6882-12-91] [PMID: 22781137]

[30] Kavitha N, Noordin R, Kit-Lam C, Sasidharan S. Real time anti-*Toxoplasma gondii* activity of an
      active fraction of *Eurycoma longifolia* root studied by *in situ* scanning and transmission electron
      microscopy. Molecules 2012; 17(8): 9207-19.
      [http://dx.doi.org/10.3390/molecules17089207] [PMID: 22858841]

[31] Pillai S, Mahmud R, Lee WC, Perumal S. Anti-parasitic activity of *myristica fragrans* houtt. essential
      oil against *toxoplasma gondii* parasite. APCBEE Procedia 2012; 2: 92-6.
      [http://dx.doi.org/10.1016/j.apcbee.2012.06.017]

[32] Yang X, Huang B, Chen J, *et al. In vitro* effects of aqueous extracts of *Astragalus membranaceus* and
      Scutellaria baicalensis GEORGI on *Toxoplasma gondii.* Parasitol Res 2012; 110(6): 2221-7.
      [http://dx.doi.org/10.1007/s00436-011-2752-2] [PMID: 22179265]

[33] Daryani A, Ebrahimzadeh MA, Sharif M, Ahmadpour E, Edalatian S, Esboei BR, *et al.*
      Anti-*Toxoplasma* activities of methanolic extract of *Sambucus nigra* (Caprifoliaceae) fruits and leaves.

Rev Biol Trop 2015; 63(1): 7-12.
[http://dx.doi.org/10.15517/rbt.v63i1.14545] [PMID: 26299111]

[34]   Happi E, Teinkela EJ, Nguemfo EL, Zambou HR, Anti GBA. *Toxoplasma gondii* activity of constituents from *Balsamocitrus camerunensis* L. (Rutaceae). Afr J Biotechnol 2014; 13(52): 4680-4.
[http://dx.doi.org/10.5897/AJB2014.14177]

[35]   Soares AM, Carvalho LP, Melo EJ, Costa HP, Vasconcelos IM, Oliveira JT. A protein extract and a cysteine protease inhibitor enriched fraction from *Jatropha curcas* seed cake have *in vitro* anti-*Toxoplasma gondii* activity. Exp Parasitol 2015; 153: 111-7.
[http://dx.doi.org/10.1016/j.exppara.2015.03.011] [PMID: 25816973]

[36]   Abugri DA, Witola WH, Jaynes JM, Toufic N. *In vitro* activity of *Sorghum bicolor* extracts, 3-deoxyanthocyanidins, against *Toxoplasma gondii*. Exp Parasitol 2016; 164: 12-9.
[http://dx.doi.org/10.1016/j.exppara.2016.02.001] [PMID: 26855040]

[37]   Nozari S, Adine M, Javadi F, *et al.* Ethanol Extracts of *Achillea millefolium* and *Hypericum perforatum* Low Anti-*Toxoplasma* Activity. J Pharmacopuncture 2016; 19(1): 70-3.
[http://dx.doi.org/10.3831/KPI.2016.19.009] [PMID: 27280052]

[38]   Javadi F, Azadmehr A, Jahanihashemi H, *et al.* Study on anti-*Toxoplasma* effects of *Myrtus communis* and *Artemisia aucheri* Boiss extracts. Int J Herb Med 2017; 5(4): 16-9.

[39]   Dégbé M, Debierre-Grockiego F, Tété-Bénissan A, *et al.* Extracts of *Tectona grandis* and *Vernonia amygdalina* have anti-*Toxoplasma* and pro-inflammatory properties *in vitro*. Parasite 2018; 25: 11.
[http://dx.doi.org/10.1051/parasite/2018014] [PMID: 29533762]

[40]   Atolani O, Oguntoye H, Areh ET, Adeyemi OS, Kambizi L. Chemical composition, anti-*Toxoplasma*, cytotoxicity, antioxidant, and anti-inflammatory potentials of *Cola gigantea* seed oil. Pharm Biol 2019; 57(1): 154-60.
[http://dx.doi.org/10.1080/13880209.2019.1577468] [PMID: 30905238]

[41]   Choi W, Jiang M, Chu J. Antiparasitic effects of *Zingiber officinale* (Ginger) extract against *Toxoplasma gondii*. J Appl Biomed 2013; 11(1): 15-26.
[http://dx.doi.org/10.2478/v10136-012-0014-y]

[42]   Dahbi A, Bellete B, Flori P, *et al.* The effect of essential oils from *Thymus broussonetii* Boiss on transmission of *Toxoplasma gondii* cysts in mice. Parasitol Res 2010; 107(1): 55-8.
[http://dx.doi.org/10.1007/s00436-010-1832-z] [PMID: 20336317]

[43]   Tavakoli Kareshk A, Keyhani A, Mahmoudvand H, *et al.* Efficacy of the *bunium persicum* (boiss) essential oil against acute toxoplasmosis in mice model. Iran J Parasitol 2015; 10(4): 625-31.
[PMID: 26811730]

[44]   Eraky MA, El-Fakahany AF, El-Sayed NM, Abou-Ouf EA, Yaseen DI. Effects of *Thymus vulgaris* ethanolic extract on chronic toxoplasmosis in a mouse model. Parasitol Res 2016; 115(7): 2863-71.
[http://dx.doi.org/10.1007/s00436-016-5041-2] [PMID: 27098159]

[45]   Gasparotto Junior A, Cosmo ML, Reis MdeP, *et al.* Effects of extracts from *Echinacea purpurea* (L) MOENCH on mice infected with different strains of *Toxoplasma gondii*. Parasitol Res 2016; 115(10): 3999-4005.
[http://dx.doi.org/10.1007/s00436-016-5167-2] [PMID: 27277433]

[46]   Mady RF, El-Hadidy W, Elachy S. Effect of *Nigella sativa* oil on experimental toxoplasmosis. Parasitol Res 2016; 115(1): 379-90.
[http://dx.doi.org/10.1007/s00436-015-4759-6] [PMID: 26446086]

[47]   Mahmoudvand H, Kareshk A, Keyhani A, Zia-Ali N, Aflatoonian M. *In vivo* evaluation of *Berberis vulgaris* extract on acute toxoplasmosis in mice. Marmara Pharm J 2017; 21(3): 558-63.
[http://dx.doi.org/10.12991/marupj.319220]

[48]   Mahmoudvand H, Beyranvand M, Nayebzadeh H, Fallahi S, Mirbadie SR, Kheirandish F, *et al.* Chemical composition, and prophylactic effects of *saturja khuzestanica* essential oil on acute

toxoplasmosis in mice. Afr J Tradit Complement Altern Med 2017; 14: 49-55.
[http://dx.doi.org/10.21010/ajtcam.v14i5.7]

[49]   de Oliveira TC, Silva DA, Rostkowska C, *et al.* *Toxoplasma gondii*: effects of *Artemisia annua* L. on susceptibility to infection in experimental models *in vitro* and *in vivo*. Exp Parasitol 2009; 122(3): 233-41.
[http://dx.doi.org/10.1016/j.exppara.2009.04.010] [PMID: 19389400]

[50]   Leesombun A, Boonmasawai S, Shimoda N, Nishikawa Y. Effects of extracts from Thai piperaceae plants against infection with *Toxoplasma gondii*. PLoS One 2016; 11(5)e0156116
[http://dx.doi.org/10.1371/journal.pone.0156116] [PMID: 27213575]

[51]   El-Tantawy NL, Soliman AF, Abdel-Magied A, *et al.* Could *Araucaria heterophylla* resin extract be used as a new treatment for toxoplasmosis? Exp Parasitol 2018; 195: 44-53.
[http://dx.doi.org/10.1016/j.exppara.2018.10.003] [PMID: 30339984]

[52]   Mirzaalizadeh B, Sharif M, Daryani A, *et al.* Effects of *Aloe vera* and *Eucalyptus* methanolic extracts on experimental toxoplasmosis *in vitro* and *in vivo*. Exp Parasitol 2018; 192: 6-11.
[http://dx.doi.org/10.1016/j.exppara.2018.07.010] [PMID: 30031121]

CHAPTER 5

Development of Antimalarial and Antileishmanial Drugs from Amazonian Biodiversity

Antônio R. Q. Gomes[1], Kelly C. O. Albuquerque[2], Heliton P. C. Brígido[1], Juliana Correa-Barbosa[1], Maria Fâni Dolabela[1,2] and Sandro Percário[2,*]

[1] *Post-Graduate Program in Pharmaceutical Innovation, Federal University of Pará, Belém, PA, Brazil*

[2] *Post-Graduate Program in Biodiversity and Biotechnology of the BIONORTE Network, Federal University of Pará, Belém, PA, Brazil*

[3] *Post-Graduate Program in Pharmaceutical Sciences, Federal University of Pará, Belém, PA, Brazil*

Abstract: The search for therapeutic alternatives for the treatment of malaria and leishmaniasis is particularly important, given the increase in parasitic resistance to available drugs, as well as the high toxicity of those drugs. In this context, the Amazon region can make an important contribution through its high biodiversity of plants, many of which are informally used by local populations for the treatment of malaria, and leishmaniasis. This chapter aims to describe the main Amazonian species used to treat malaria and leishmaniasis in Brazilian folk medicine, relating ethnobotanical results to chemical studies, evaluation of activities, and toxicity. Different studies report the treatment of malaria with plants, with the most cited species being *Aspidosperma nitidum* Benth. (Apocynaceae); *Geissospermum sericeum* (Sagot.) Benth & Hook (Apocynaceae); *Euterpe precatoria* Mart. (Arecaceae); *Persea americana* Mill (Lauraceae); *Bertholletia excelsa* Bonpl (Lecythidaceae); *Portulaca pilosa* L. (Portulaceae); *Ampelozizyphus amazonicus* Ducke (Rhamnaceae). Additionally, traditional Amazonian populations use plants for the treatment of wounds, a clinical aspect associated with leishmaniasis, with the most cited genus being *Copaiba* and *Jatropha*. The antileishmanial activity of copaiba oil has been demonstrated, and it seems that this activity is related to terpenes. Another genus that deserves attention is *Musa*, used for the treatment of severe wounds. The leishmanicidal activity of triterpenes isolated from *Musa paradisiaca* and its anacardic acid and synthetic derivatives, which have been used against *Leishmania infantum chagasi*, was also tested. In summary, several isolated compounds of plants used in traditional Amazonian medicine are promising as antimalarial and antileishmanial drugs.

---

[*] **Corresponding author Sandro Percário**: Post-Graduate Program in Biodiversity and Biotechnology of the BIONORTE Network, Federal University of Pará, Belém, PA, Brazil;
E-mails: spercario49@gmail.com and percario@ufpa.br

Atta-ur-Rahman, *FRS* (Ed.)

Keywords: Amazon, *Aspidosperma nitidum, Bertholletia excelsa, Copaiba, Euterpe precatoria, Geissospermum sericeum, Jatropha gossypiifolia,* Leishmaniasis, Malaria, Medicinal plants, *Musa parasidiaca, Persea americana.*

INTRODUCTION

The recognition of the environmental limits of the modern development model has imposed the need for new forms of global governance upon the planetary environment, requiring proposals for sustainable development that oppose the worsening of environmental degradation and biodiversity loss [1].

In Brazil, the Amazon region and its people have been threatened by short-sighted, profit-driven economic interests, driving the increase in deforestation in an increasingly chaotic way. According to the Real-Time Legal Amazon Deforestation Detection System (DETER), deforestation alerts were recorded in an area of 4,219.3 square kilometers in 2018, and in 2019, 9,165.6 square kilometers of forest were deforested – more than double the area recorded in the previous year.

This accelerated deforestation will probably result in the extinction of many plant species, which will have negative impacts on the culture of the use of medicinal plants by the peoples of the Amazon. In 2005, it was estimated that about 180 indigenous peoples (approximately 208,000 individuals) lived in the Amazon, in addition to 357 remaining *quilombola* (maroon) communities and thousands of rubber tapper, riverside or babassu communities [2]. In fact, in addition to its biodiversity in terms of plant and animal species, due to the different ethnicities of its peoples, the Amazon also displays a wide spectrum of cultural diversity.

As a result of this fact, another important issue arises, which is the understanding of the process of occupation of the Amazon and the impact on the health of indigenous peoples and people who settled in the region. In this sense, this process stimulated the occurrence of several epidemics and created an asymmetry in the access to health services. For example, in metropolises, such as Belém and Manaus, health services are structured, while in remote locations within the forest, due to the great difficulty of access, the only therapeutic alternative available to treat diseases has often been the use of medicinal plants [3].

Among these diseases, malaria has been affecting the Amazonian people for centuries. A study conducted in 1885 showed that the Amazon was already plagued by the disease, and the possible explanation for this fact results from the intense migration that occurred to this region in the nineteenth century, resulting from rubber extraction activities and the construction of the Madeira-Mamoré railroad. In this scenario, many immigrants ended up dying from malaria – which is considered the second major epidemic witnessed by Osvaldo Cruz and Carlos Chagas [4]. At the same time, as opposed to the high mortality experienced by immigrants, the riverside populations survived in this hostile environment due to their ancient knowledge of many native and exotic plant species to treat malaria and its symptoms, and this medicinal information was orally transmitted from one generation to another [5].

Deforestation in the Amazon is still a serious medical and public health problem, as it creates conditions for the development of several tropical endemic diseases, such as American Tegumentary Leishmaniasis (ATL). During deforestation, the rodent population migrates to other areas in search of natural shelters, and phlebotomic fauna, which previously engaged in hematophagy using these small mammals, begin to seek out humans for this purpose [6], thus transmitting the etiological agents of various diseases, such as malaria and leishmaniasis, among others. In this context, plant species that have historically been used in Amazonian folk medicine as healing agents and for wound treatment have shown promise for the treatment of tegumentary leishmaniasis.

This chapter aims to describe the main Amazonian species that are used in folk medicine for the treatment of malaria and leishmaniasis, relating ethnobotanical results to chemical studies, evaluation of activities, and toxicity. Initially, a search was performed for ethnobotanical studies available in different databases, and species with claims of use for malaria and leishmaniasis, for wound treatment, or as healing agents were selected.

Medicinal Plants used for the Treatment of Malaria and Leishmaniasis

Ethnobotanical studies have already been conducted in some regions of the Amazon, but in other regions, there is a lack of scientific studies aimed at describing the uses of plants for medicinal purposes. A range of factors contributes to create difficulties in conducting such studies, such as the large dimension of the territory of the Brazilian Amazon, and difficulties in moving across certain regions due to the lack of connecting roads, implying the need for air transport associated with river transport, which greatly increases the cost of a study. Another important factor is the reduced number of researchers in the ethnobotany field that reside in this region, in addition to scarce funding for these studies because of public policies related to research funding in Brazil.

Notwithstanding, among the studies already published, there are reports of popular use of plants for the treatment of malaria for more than 100 plant species of the region. The most cited families were Apocynaceae, Araceae, Arecaceae, Euphobiaceae, Fabaceae, Menispermaceae, Rhamnaceae and Simaroubacea. Table **1** summarizes the species that were mentioned in the available literature, with the species with the highest number of citations being *Aspidosperma nitidum*; *Geissospermum sericeum*; *Euterpe precatoria*; *Persea americana; Bertholletia excelsa; Portulaca pilosa* and *Ampelozizyphus amazonicus*.

In addition to *A. nitidum* (Fig. **1**), antimalarial property has been attributed to *A. excelsum* [7, 8] (Table **1**), and the most recent botanical study considers the species *A. nitidum* a synonym of *A. excelsum* [9]. Other species of *Aspidosperma* are used for the treatment of malaria in the Amazon region [10 - 12] (Table **1**) and in other regions of Brazil [13, 14]. Another genus belonging to the family Apocynaceae, widely used in traditional Brazilian medicine for the treatment of malaria is *Geissospermum*, in particular the species *G. sericeum* [15 - 18] and *G. vellosii* [19, 20], both being used in the form of infusions of their barks for this medicinal purpose.

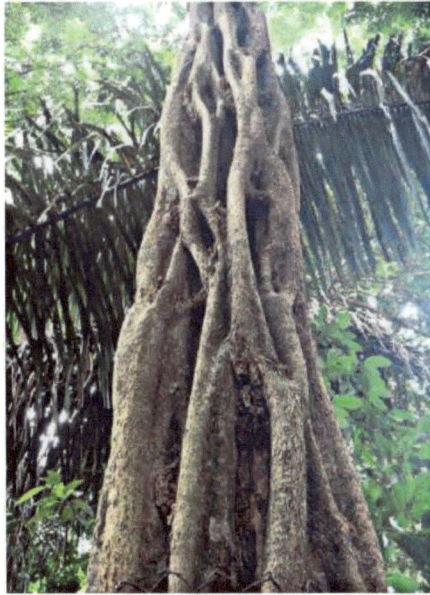

Fig. (1). *Aspidosperma nitidum* (courtesy of Dr. Pedro Pompei Filizzola Oliva and Museu Paraense Emilio Goeldi).

Euterpe precatoria (Fig. **2**), known in Brazil as *açaizeiro, açaí-do-amazonas* or *açaí-solitário*, is a species native to the Amazon, with great importance as an Amazonian food source and for popular medicine. In addition to the claim of use for the treatment of malaria [11, 12, 14] (Table **1**), this species is used to treat muscle pain, chest pain, snake bites, and in the treatment of flu, along with some very bizarre claims, such as for the hair to grow well and very black, and to prevent pregnant women from losing hair [21].

Likewise, several medicinal properties are attributed to *Persea americana* (Fig. **3**) - known as the avocado tree - from which teas and macerations are made from its leaves and seeds [22]. Ethnobotanical studies show that in Central America, the infusion of toasted leaves from *P. americana* is commonly used for malaria treatment [14].

Fig. (2). *Euterpe precatoria Mart.* (courtesy of Dr. Pedro Pompei Filizzola Oliva and Museu Paraense Emilio Goeldi).

Fig. (3). *Persea americana* (courtesy of Dr. Pedro Pompei Filizzola Oliva and Museu Paraense Emilio Goeldi).

Bertholletia excelsa (Fig. **4**), known as the Brazil nut tree, is used to make tea and juices for topical use and macerations for internal use from its stem barks and fruits [22]. A study conducted in Acre reports the use of tea from this plant for the treatment of malaria [23], while in Manaus, the use of tea from the leaves has been reported [24]. In the Amazon region, decoctions of the bark of the species are also used for this purpose [14].

Fig. (4). *Bertholletia excelsa* Bonpl (courtesy of Ms. Paula Maria Correa de Oliveira and Dr. Marlia Regina Coelho Ferreira).

Ampelozizyphus amazonicus is popularly known as *saracuramirá* and is widely distributed throughout South America, being found in Amazonian territories of Brazil, Venezuela, Colombia, Peru, and Ecuador [25]. In the Amazon region, decoction obtained from the root of *saracuramirá* is used in malaria prevention [26, 27] and treatment [28 - 30]. *In vivo* and *in vitro* antimalarial activity has already been demonstrated against the sporozoite form of *Plasmodium* [31], supporting the popular claim made for this species in the Amazon region.

Table 1. Ethnobotanical studies of plants popularly used for the treatment of malaria

Family	Species	References
Annonaceae	*Guatteria guianensis* (Aubl.)	Kffuri *et al.*, 2016 [11] Kffuri *et al.*, 2019 [12]
Apocynaceae	*Aspidosperma excelsum* Benth.	Vásquez *et al.*, 2014 [22]
	Aspidosperma nitidum Benth.	Altschul, 1973 [32] Brandão *et al.*, 2020 [33] Milliken & Albert, 1996 [34] Milliken, 1997 [18] Scudeller *et al.*, 2009 [35] Tomchinsky *et al.*, 2017 [14]
	Aspidosperma rigidum Rusby	Oliveira *et al.*, 2011 [10]
	Aspidosperma schultesii Woodson	Kffuri *et al.*, 2016 [11] Kffuri *et al.*, 2019 [12]
	Aspidosperma spp.	Killeen *et al.*, 1993 [36] Veiga & Scudeller, 2015 [13] Tomchinsky *et al.*, 2017 [14]
	Geissospermum argentum Woodson	Oliveira *et al.*, 2011 [10]
	Geissospermum sericeum Sagot	Le Cointe, 1947 [37] Correa, 1975 [15] Cruz & da Silva, 1979 [16] Balbach, 1980 [17] Milliken, 1997 [18]
	Himatanthus articulatus Vahl.	Milliken, 1997 [18]

(Table 1) cont.....

Family	Species	References
Araceae	*Heteropsis* sp., *H. tenuispadix* G.S. Bunting	Frausin *et al.*, 2015 [38] Kffuri *et al.*, 2016 [11]
	Attalea maripa Aubl.	Kffuri *et al.*, 2016 [11] Kffuri *et al.*, 2019 [12]
	Euterpe catinga Wallace	Kffuri *et al.*, 2016 [11] Kffuri *et al.*, 2019 [12]
	Euterpe oleracea Mart.	Brandão *et al.*, 1992 [39] Vigneron *et al.*, 2005 [40] Tomchinsky *et al.*, 2017 [14]
	Euterpe precatoria Mart.	Deharo *et al.*, 2001 [41] Hidalgo, 2003 [42] Bertani *et al.*, 2005 [43] Balslev *et al.*, 2008 [44] Hajdu & Hohmann, 2012 [45] Vásquez *et al.*, 2014 [22] Frausin *et al.*, 2015 [38] Veiga & Scudeller, 2015 [13] Kffuri *et al.*, 2016 [11] Tomchinsky *et al.*, 2017 [14] Kffuri *et al.*, 2019 [12]
Arecaceae	*Iriartea deltoidea* Ruiz & Pav.	Kffuri *et al.*, 2016 [11] Kffuri *et al.*, 2019 [12]
	Vernonia condensata Baker	Milliken, 1997 [18] de Oliveira *et al.*, 2016 [46]
Bignoniaceae	*Jacaranda copaia* Aubl.	Kffuri *et al.*, 2016 [11] Kffuri *et al.*, 2019 [12]
Caricaceae	*Carica papaya* L.	Milliken, 1997 [18] de Oliveira *et al.*, 2016 [46]
Celastraceae	*Maytenus guianensis* Klotzch ex Reissek	Oliveira *et al.*, 2015 [47] Veiga & Scudeller, 2015 [13] Cajaiba *et al.*, 2016 [48]
Compositae	*Acanthospermum australe* Loefl.	Braga, 1960 [49] Correa, 1975 [15] Cruz & da Silva, 1979 [16]
Convolvulaceae	*Bonamia ferruginea* Choisy	Paes & Mendonça, 2008 [50] Veiga & Scudeller, 2015 [13]
Costaceae	*Costus spicatus* Jacq.	Hidalgo, 2003 [42] de Oliveira *et al.*, 2016 [46]
Cucurbitaceae	*Momordica charantia* L.	Milliken, 1997 [18] de Oliveira *et al.*, 2016 [46]

(Table 1) cont.....

Family	Species	References
Euphorbiaceae	*Croton cajucara* Benth.	Milliken, 1997 [18] Veiga & Scudeller, 2015 [13]
	Jatropha gossypiifolia L.	Coutinho *et al.*, 2002 [51] Vásquez *et al.*, 2014 [22]
Fabaceae	*Hymenaea courbaril* L.	Oliveira *et al.*, 2015 [47] Vásquez *et al.*, 2014 [22]
	Monopteryx uaucu Spruce ex Benth.	Kffuri *et al.*, 2016 [11] Kffuri *et al.*, 2019 [12]
	Ormosia discolor Spruce ex Benth.	Kffuri *et al.*, 2016 [11] Kffuri *et al.*, 2019 [12]
	Swartzia argentea Spruce ex Benth.	Kffuri *et al.*, 2016 [11] Kffuri *et al.*, 2019 [12]
Lauraceae	*Persea americana* Mill.	Milliken, 1997 [18] Hidalgo, 2003 [42] Blair & Madrigal, 2005 [52] Coelho-Ferreira, 2009 [53] Oliveira, 2011 [10] Tomchinsky *et al.*, 2017 [14]
Lecythidaceae	*Bertholletia excelsa* Bonpl.	Brandão *et al.*, 1992 [39] Hidalgo, 2003 [42] Coelho-Ferreira, 2009 [53] Tomchinsky *et al.*, 2017 [14]
Menispermaceae	*Abuta* sp.	Frausin *et al.*, 2015 [38] Arevalo, 1994 [54]
Menispermaceae	*Cissampelos ovalifolia* DC.	Milliken, 1997 [18] de Oliveira *et al.*, 2016 [46]
Nyctaginaceae	*Boerhavia hirsuta* Willd.	Delorme & Miola, 1979 [55] Neves, 1980 [56]
Piperaceae	*Piper* sp., *Piper cernuum* Vell.	Kffuri *et al.*, 2016 [11]
Portulaceae	*Portulaca pilosa*	Neves, 1980 [56] da Silva *et al.*, 1998 [57] Souza, 2010 [58] Veiga & Scudeller, 2015 [13] Ferreira *et al.*, 2015 [23] Pinheiro, 2018 [24]

(Table 1) cont.....

Family	Species	References
Rhamnaceae	*Ampelozizyphus amazonicus* Ducke	Neves, 1980 [56] Paulino-Filho, 1979 [59] Brandão *et al.*, 1992 [39] Milliken, 1997 [18] Hidalgo, 2003 [42] Oliveira *et al.*, 2011 [10] Vásquez *et al.*, 2014 [22] Kffuri *et al.*, 2016 [11] Tomchinsky *et al.*, 2017 [14] Kffuri *et al.*, 2019 [12]
Rubiaceae	*Sabicea amazonenses* Wernham	Kffuri *et al.*, 2016 [11] Kffuri *et al.*, 2019 [12]
Simaroubaceae	*Simaba cedron* Planch.	Altschul, 1973 [32] Oliveira *et al.*, 2011 [10] Frausin *et al.*, 2015 [38]
Strelitziaceae	*Phenakospermum guianensis* Rich.	Kffuri *et al.*, 2016 [11] Kffuri *et al.*, 2019 [12]
Verbenaceae	*Stachytarpheta cayennensis* Rich.	Milliken, 1997 [18] Oliveira *et al.*, 2003 [60]

Leishmaniasis is a parasitic disease caused by *Leishmania*, which infects the vertebrate host through the bite of female vectors of the genera *Lutzomyia* [61, 62]. Different studies conducted in the Amazon region have demonstrated the popular use of plants in the treatment of wounds and leishmaniasis. Table **2** summarizes the species cited in the studies conducted in the Amazon region.

Portulaca pilosa (Fig. **5**) and *Aspidosperma* (Fig. **1**) were also cited for the treatment of malaria and wound healing [22, 23, 63, 64] (Tables **1-2**, respectively). *Copaíba* oil, obtained from different species of *Copaifera* (Fig. **6**), is widely used for wound treatment and healing [65 - 67]; there have also been reports of the use of bark and leaves to produce the healing effect [68] (Table **2**).

Another genus with a popular claim of wound healing is *Jatropha* (Fig. **7**; Table **2**), and several studies have evaluated its healing effects [69 - 72]. In addition to medicinal use, this species has ornamental utility [73]. Preparations of *J. gossypiifolia* are also used in religious rituals [74, 75] and for the construction of living fences or hedges that are used against the spreading of fires [73, 76]. Other uses reported for this species include its insecticide action [77], and the use of seed oil in the preparation of paints, soaps, lubricants, and fuel for diesel engines and for lighting [73, 76].

Fig. (5). *Portulaca pilosa* (courtesy of Dr. Pedro Glecio Costa Lima and Dr. Marlia Regina Coelho Ferreira).

Fig. (6). *Copaifera reticulada* (courtesy of Dr. Pedro Pompei Filizzola Oliva and Museu Paraense Emilio Goeldi).

Fig. (7). *Jatropha curcas* (courtesy of Dr. Pedro Pompei Filizzola Oliva and Museu Paraense Emilio Goeldi).

Musa acuminata and *M. paradisiaca* (Fig. **8**) are species of banana trees that have several claims of popular use, including as a sedative for toothache, healing of surgical wounds from tooth extraction, gastric ulcers, hypoglycemia, as an antidote to snake bites, and diarrhea, among others [78]. The parts of the plant that can be used for medicinal purposes are flowers, roots, fruits, and latex, and are applied topically or internally.

Table 2. Ethnobotanical studies of plants used for the popular treatment of wounds and leishmaniasis

Family	Species	References
Anacardiaceae	*Schinus terebinthifolius* Raddi.	Silva *et al.*, 2011 [79]
Apocynaceae	*Aspidosperma excelsum* Benth	Vásquez *et al.*, 2014 [22]

(Table 2) cont.....

Family	Species	References
Bignoniaceae	*Crescentia cujete* var.	Sarquis *et al.*, 2019 [80]
	Fridericia chica Bonpl.	Vásquez *et al.*, 2014 [22]
Boraginaceae	*Symphytum officinale* L.	Cajaiba *et al.*, 2016 [48]
Celastraceae	*Maytenus guianensis* Klotzch ex Reissek	de Oliveira *et al.*, 2015 [47] Veiga & Scudeller, 2015 [13] Cajaiba *et al.*, 2016 [48]
Chenopodiaceae	*Chenopodium ambrosioides* L.	Cajaiba *et al.*, 2016 [48] Scudeller *et al.*, 2009 [35]
Fabaceae	*Copaifera* sp.	Santana *et al.*, 2014 [66] Cavalcante *et al.*, 2017 [68]
	Copaifera langsdorffii	Cavalcante *et al.*, 2017 [68]
	Copaifera marti	Roman & Santos, 2006 [65]
	Copaifera pubiflora Benth.	Oliveira *et al.*, 2019 [67]
Euphorbiaceae	*Jatropha gossypiifolia* L.	Coutinho *et al.*, 2002 [51] Matos, 2004 [73] Aquino *et al.*, 2006 [69] Maia *et al.*, 2006 [70] Santos *et al.*, 2006 [71] Vale *et al.*, 2006 [72] Vásquez *et al.*, 2014 [22]
	Jatropha curcas L.	Vásquez *et al.*, 2014 [22] Leão *et al.*, 2007 [81]
Fabaceae	*Libidibia ferrea* Mart. ex Tul.	Sarquis *et al.*, 2019 [80]
	Manihot esculenta Crantz	Vásquez *et al.*, 2014 [22]
Lamiaceae	*Plectranthus barbatus* Andrews	Vásquez *et al.*, 2014 [22]
Meliaceae	*Carapa guianensis* Aubl.	Cajaiba *et al.*, 2016 [48]
Musaceae	*Musa acuminata* Colla	Vásquez *et al.*, 2014 [22]
	Musa paradisiaca L.	Vásquez *et al.*, 2014 [22]
Myrtacea	*Eugenia punicifolia* Kunth	Vásquez *et al.*, 2014 [22]
Plantaginacea	*Scoparia dulcis* L.	Vásquez *et al.*, 2014 [22]
Portulaceae	*Portulaca pilosa*	Da silva *et al.*, 1998 [57] Mors *et al.*, 2000 [82] Revilla, 2002 [83] Alves *et al.*, 2006 [84] Barata *et al.*, 2009 [85] Oak, 2015 [86] Fereira *et al.*, 2015 [23] Nunes, 2016 [63] Barros *et al.*, 2017 [64]

(Table 2) cont.....

Family	Species	References
Simaroubaceae	*Quassia amara* L.	Botsaris, 2007 [87]
	Simarouba amara Aubl.	Botsaris, 2007 [87]

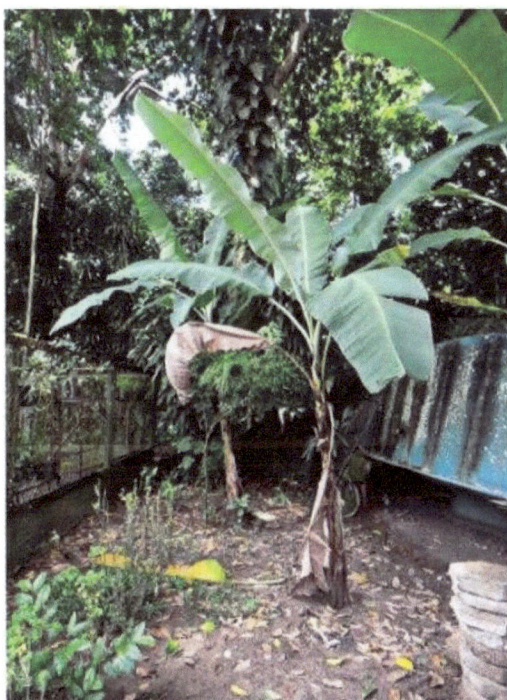

Fig. (8). *Musa parasidiaca* (courtesy of Dr. Pedro Pompei Filizzola Oliva and Museu Paraense Emilio Goeldi).

Chemical Studies and Evaluation of Biological Activities of Species that are more Frequently Cited in the Literature

Several studies have shown that different plant species are used in traditional medicine for the treatment of malaria and leishmaniasis. However, some species deserve special attention, and among these are *Portulaca pilosa*, *Aspidosperma* and Apocynaceae (*Geissospermum* and *Himatanthus*).

Phytochemical studies conducted from the ethanol extract of the aerial parts of *P. pilosa* show the isolation of diterpenes, such as pilosanone A and C (Fig. **9**) [88]. In another study, from the ethyl acetate fraction of the roots of *P. pilosa*, three clerodane diterpenes were isolated: pilosanol A, B and C (Fig. **9**) [89].

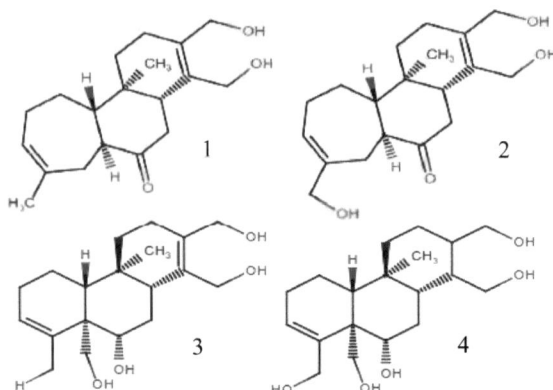

Fig. (9). Diterpenes isolated from *Portulaca pilosa*. **Legend**: *(1) Pilosanone A; (2) Pilosanone B; (3) Pilosanol A; (4) Pilosanol B.*

The ethanol extract obtained from the aerial parts of *P. pilosa* underwent a study *in vitro* against promastigote and amastigote forms of *Leishmania amazonensis*, but no promising activity has been demonstrated [90]. Another study evaluated the healing activity in surgical lesions of Wistar rats using gel and propylene glycol extracts from *P. pilosa* (150mg/kg), and the histological analysis of the lesions showed that the extract modulated the inflammatory response of the tissue, stimulated angiogenesis and fibroblast proliferation. In groups treated with *P. pilosa,* healing was better than the negative control, and a better pattern of organization of the epidermis and dermis was observed, in a mild inflammatory process, with fibroblast proliferation and increased collagen fiber formation. The topical anti-inflammatory activity is probably related to gallic acid, the phytochemical marker of this species [64].

An *in vitro* study carried out against *Plasmodium* showed that the ethanol extract obtained from aerial parts of *P. pilosa* was active and presented low cytotoxicity to macrophages, and high selectivity was observed. Further studies need to be conducted to verify antimalarial activity using *in vivo* models and to identify which compound is involved in this activity [33].

Another important species, *Aspidosperma nitidum (*synonym *A. excelsum),* was subjected to chemical studies, and the following alkaloids were isolated: 11-methoxy tubotaiwine (Fig. **10-1**), compactinervine (Fig. **10-2**), N-acetyl aspidos-permidine (Fig. **10-3**), O-desmethyl-aspidospermidine (Fig. **10-4**), aricine (Fig. **10-5**), yohimbine (Fig. **10-6**), tetrahydrosecamine (Fig. **10-7**), 16-desmethox--carboxyl-tetrahydrosecamine (Fig. **10-8**), didesmethoxy-carboxyl- tetrahydro-secamine (Fig. **10-9**), O-acetyl yohimbine (Fig. **10-10**) yohimbine, ocryl fuanine [91] (Fig. **10-11**), excelsinine [92] (Fig. **10-12**), 10-methoxygeissoschizol (Fig.

10-13), 10-methoxyyohimbine (Fig. **10-14**), and 10-methoxy-4-methylgeissos-chizol [93] (Fig. **10-15**). O-acetyl yohimbine and 10-methoxycorynanthine (Fig. **10-16**) were isolated from the root bark of *A. excelsum* [94].

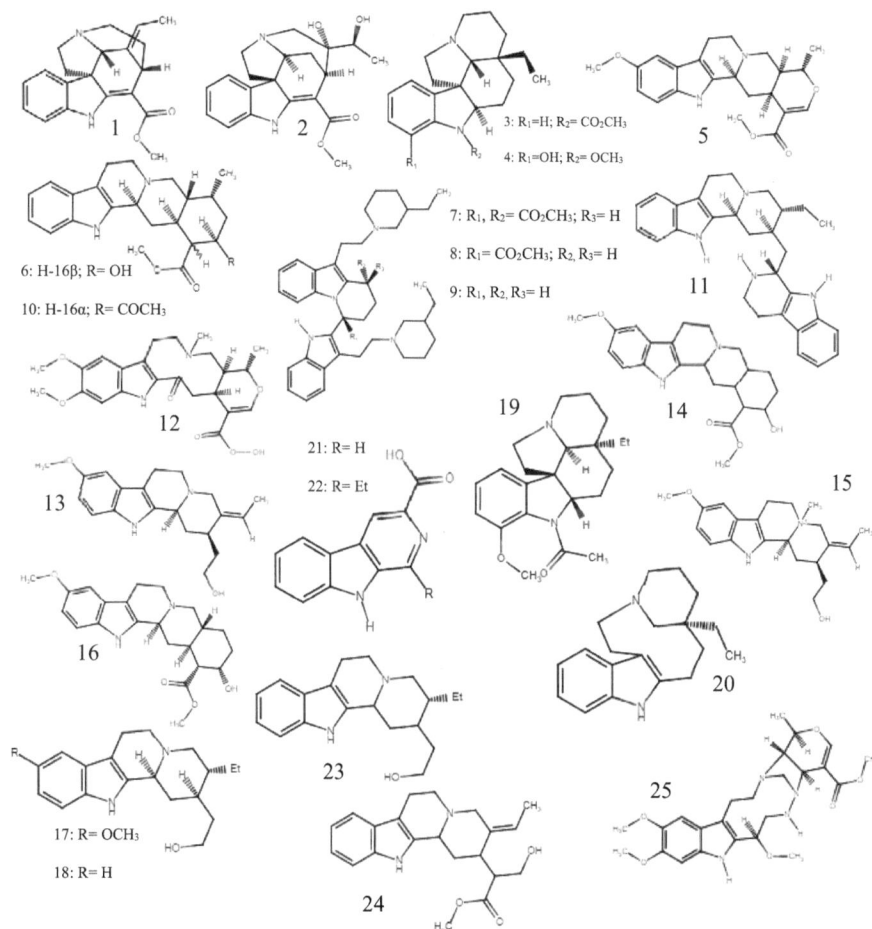

Fig. (10). Chemical structure of compounds occurring in *Aspidosperma excelsum* and *Aspidosperma nitidum*. **Legend:** 11-methoxytubotaiwine (1), compactinervine (2), N-acetyl aspidospermidine (3), O-desmethyl aspidospermidine (4), aricine (5), yohimbine (6), tetrahydrosecamine (7), 16-desmemethoxy-carboxyl-tetrahydrosecamine (8), didesmethoxy-carboxyl tetrahydrosecamine (9), O-acetyl yohimbine (10), ocryl fuanine (11), excelsinine (12), 10- methoxygeissoschizol (13), 10-methoxyyohimbine (14), 10-methoxy-4-methylgeissoschizol (15), 10-methoxycorynanthine (16), 10-methoxy-dihydro-corynantheol (17), corynantheol (18), aspidospermine (19), quebrachamine (20), carboxylic harman acid (21), 3-carboxylic ethylharman (22), dihydrocorynantheol (23), dehydrositsiriquine (24), and braznitidumine (25).

Other phytochemical studies of *A. nitidum* have also isolated 10-methoxy-dihydro-corynantheol (Fig. **10**-17), corynantheol [95] (Fig. **10**-18), aspidospermine (Fig. **10**-19), quebrachamine (Fig. **10**-20), yohimbine [96] (Fig. **10**-6), carboxylic harman acid (Fig. **10**-21), 3- carboxylic ethylharman [97] (Fig. **10**-22), dihydrocorynantheol (Fig. **10**-23), dehydrositsiriquine [98] (Fig. **10**-24), and braznitidumine [98] (Fig. **10**-25).

The antiplasmodial activity of the ethanol extract obtained from the bark of *A. nitidum* proved to be active against clone of *Plasmodium falciparum* resistant to chloroquine and fractionation led to the obtainment of a more active fraction (fraction of alkaloids). In all doses used (125-500 mg/kg), on the 5th day of infection, a significant reduction in parasitemia was observed in mice infected with *Plasmodium berghei* [99].

In relation to leishmanicidal activity, the ethanol extract obtained from barks of *A. nitidum* proved to be active against promastigote forms of *L. amazonensis*, but the fraction of alkaloids seems to be less promising. This activity was suggested to be related to the synergism between alkaloids and other compounds presented in this species [100].

Another species widely used in the Brazilian Amazon is *Geissospermum sericeum*. In the phytochemical prospection of the ethanol extract obtained from *G. sericeum*, alkaloids, flavonoids, tannins and saponin were detected [101]. From the extracts obtained from the stem barks of *G. sericeum,* the following alkaloids were isolated: geissospermine [102] (Fig. **11**-1), geissoschizoline (Fig. **11**-2), geissoschizoline N4-oxide (Fig. **11**-3), 1,2 dihydrogeissoschizoline (Fig. **11**-4), and flavopereirine [103] (Fig. **11**-5).

The aqueous extract obtained from stem barks of *G. sericeum* proved inactive against *Plasmodium berghei* [104], but the hydromethanic extract and isolated alkaloids were active in an *in vitro* study against clones of *P. falciparum* resistant to chloroquine (K1), with the alkaloid flavopereirine being considered more promising [103]. Similarly, flavopereirine was very promising against *L. amazonensis*, as well as was the fraction of alkaloids [105].

Antimalarial and antileishmanial activities of *A. nitidum* and *G. sericeum* have been related to alkaloids. Some isolated alkaloids present in these species have already undergone *in vitro* studies to evaluate antiparasitic activities (Table **3**).

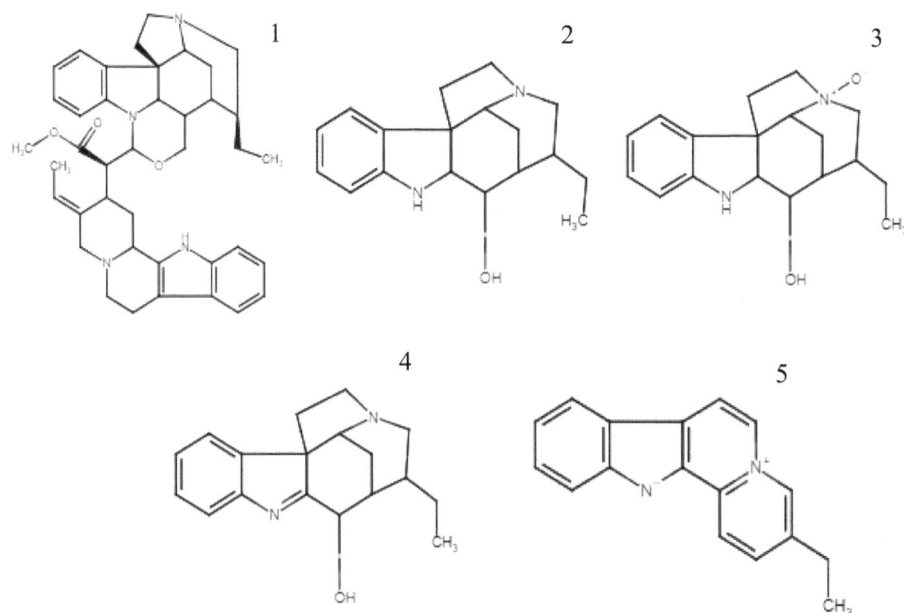

Fig. (11). Chemical structure of compounds isolated from *Geissospermum sericeum*.
Legend: geissospermine **(1)**, geissoschizoline **(2)**, geissoschizoline N4-oxide **(3)**; 1,2 dihydrogeissoschizoline **(4)**, and flavopereirine **(5)**.

Table 3. Antimalarial and antileishmanial activities of alkaloids isolated from *Aspidosperma nitidum* and *Geissospermum sericeum*.

Alkaloids	Antimalarial	Antileishmania	References
O-demethylaspidospermidine	Active against *P. falciparum* chloroquine-resistant	Active against *L. infantum* (CI_{50} = 7.7 0 µg/mL)	Reina *et al.*, 2012 [106]
Aricine	Active against *P. falciparum* chloroquine-resistant (IC_{50} 0.69 µM)	NE	Passemar *et al.*, 2011 [107]
Yohimbine	Active (W2, IC_{50} 14.35 ± 2.77)	NE	do Nascimento *et al.*, 2019 [108]
Aspidospermine	Active against *P. falciparum* chloroquine-resistant and -sensitive (IC_{50} 3.8 ± 0.7 and 4.6 ± 0.5 µM, respectively)	NE	Mitaine-Offer *et al.*, 2002 [109]
Quebrachamine	NE	NE	Saxton, 1996 [110]
Carboxylic harman acid	Inactive	NE	Coutinho *et al.*, 2013 [106]

(Table 3) cont.....

Alkaloids	Antimalarial	Antileishmania	References
Braznitidumine	Active against *P. falciparum* (IC$_{50}$ 8.3±1.6µg/mL)	NE	Coutinho *et al.*, 2013 [8]
Geissospermine	Active against *P. falciparum* (D10) sensitive to chloroquine (IC$_{50}$ 5.02 ± 0.74 µM)	NE	Mbeunkui *et al.*, 2012 [19]
Geissoschizoline	Inactive	NE	Steele *et al.*, 2002 [103]
Geissoschizoline N4 -oxide	Inactive	NE	Steele *et al.*, 2002 [103]
1,2-de-hydrogeissoschizoline	Inactive against K1 and strains of *P. falciparum* T9-96 (IC$_{50}$ 27.26 ± 10.9 and 35.37 ± 2.36 µM, respectively)	NE	Steele *et al.*, 2002 [103]
Flavopereirine	Active against *P. falciparum*	Active against *L. amazonensis*	Steele *et al.*, 2002 [103] Silva *et al.*, 2019 [105]

Legend: NE- unstudied; IC$_{50}$- inhibitory concentration 50%; Clones of *Plasmodium falciparum*: W2- resistant to chloroquine; K1- chloroquine resistant D10- chloroquine-sensitive; *L. infantum-Leishmania infantum.*

Nevertheless, it is observed that most antiparasitic studies of alkaloids isolated from *A. nitidum* and *G. sericeum* were evaluated only against malaria (Table **3**). Only O-demethylaspidospermidine and flavopereirine alkaloids were evaluated against *Leishmania*, where the former proved to be active against *L. chagasi* [106], and the latter against *L. amazonensis* [105].

As for antimalarial activity, it is observed that the alkaloids O-demethylaspidospermidine, aricine, yohimbine, aspidospermine, quebrachamine, braznitidumine, geissospermine, 1,2-de-hydrogeissoschizoline and flavopereirine were active against strains of *P. falciparum* in several studies [8, 19, 103, 106 - 109].

From the leaves of *Persea americana*, one previously undescribed flavonol glycoside (Fig. **12**-1) together with ten known flavonoids (Fig. **12**-2-11), four megastigmane glycosides (Fig. **12**-12–15) and two lignans (Fig. **12**-16–17) were isolated [111].

The aqueous extract of *P. americana* was active against clones of *P. falciparum* sensitive to chloroquine (3d7; IC$_{50}$= 9.93 ± 0.86 µg/mL) and chloroquine resistant (W2; IC$_{50}$= 34.20 ± 5.80 µg/mL), presenting high selectivity for clone 3d7 [112] (Selective Index >10.1). The methanol extract of *P. americana* leaves possesses significant antimalarial activity against *P. berghei*-infected mice (p<0.05). The results also show that the extract exhibits excellent hematopoietic properties by reversing and restoring the altered plasmodium-induced changes and hematological indexes [113].

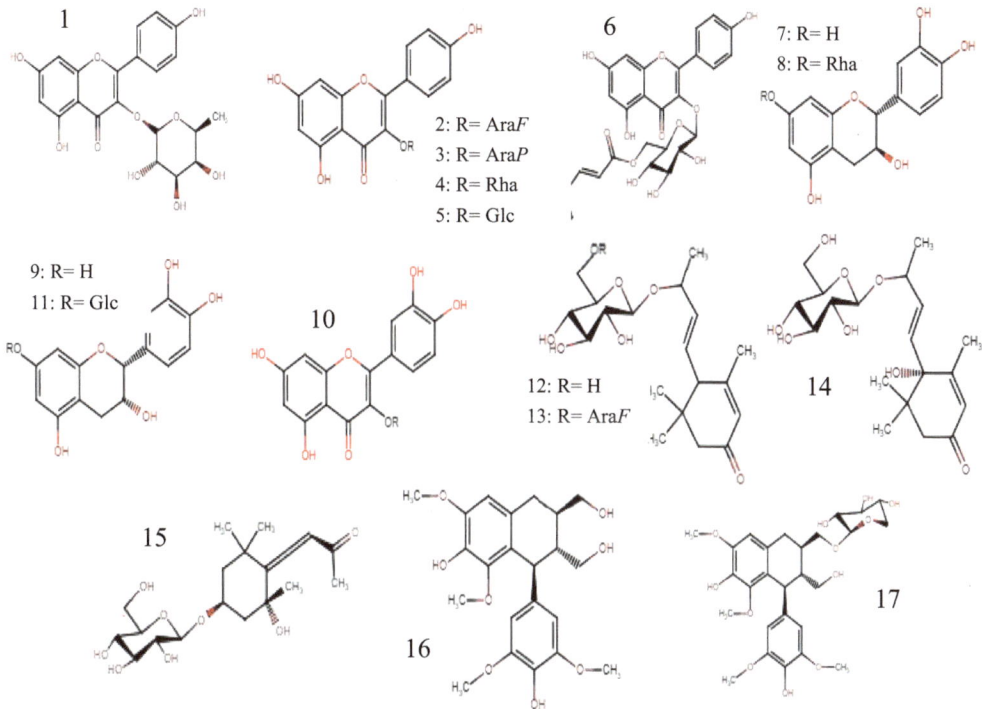

Fig. (12). Chemical structure of compounds isolated from *Persea americana*.
Legend: glycoside flavonol (1), juglanin (2), juglalin (3), afzelin (4) astragaline (5), trans-tiliroside (6), quercetin (7), quercitrin (8), catechin (9), epicatechin (10), senecin (11), (6R,9R)-3-oxo-alpha-ionol- 9-O-d-glucopyranoside (12), ficumegasoside (13), (6S,9R)-roseoside (14), icariside B1 (15), (+)-lyoniresinol (16), and (+)-isolariciresinol 9-O-D-xylopyranoside (17).

In relation to the isolated compounds of this species and its antiplasmodial activity, after extensive review, it was found that five isolated compounds were active *in vitro* against *P. falciparum* clones sensitive (D6) and resistant (W2) to chloroquine, based on the plasmodial LDH activity assay. The compound 2S4S-1,2,4-trihydroxyheptadec-16-ene was the most active against both plasmodium strains [114] (IC$_{50}$ = 1.6 and 1.4 µg/mL for the D6 clone, respectively, and 2.1 and 1.4 µg/mL for the W2 clone, respectively.

Moreover, the dichloromethane extract obtained from fruits of the species *P. americana* showed moderate activity against promastigote forms of *Leishmania donovani*. However, the activity-guided fractionation of the above extract led to the isolation of two acetogenins (5*E*,12*Z*,15*Z*)-2-hydroxy-4-oxohenicosa-5,-2,15-triene-5-1-yl acetate and (2*E*,5*E*,12*Z*,15*Z*)-1-hydroxyphone-2,5,12,15-triene-4-one, which showed good antileishmanial activity [115]. From this species, compounds were isolated with several flexible hydrophobic ligands, geranylgeraniol and C17 fatty alcohol derivatives, which showed selective docking for *Trypanosoma cruzi* trypanothione reductase [116]. Thus, *P. americana* seems to be promising both as an antimalarial and antileishmanial agent, and perhaps acetogenins are involved in these activities.

Another species of great importance as a food source in the Amazon region and that is used in folk medicine is *Euterpe precatoria*. From the root powder and leaf splint, hexane, ethyl acetate, and methanolic extracts were obtained. From these extracts it was isolated β-sitosterol (Fig. **13**-1) and stigmasterol (Fig. **13**-2), stigmast-4-en-6b-ol-3-one (Fig. **13**-3), acid *P*-hydroxybenzoic (Fig. **13**-4), 3b-*O*-b-D- glucopyranosyl sitosterol (Fig. **13**-5), palmitate β -sitosterile (Fig. **13**-6), mixture of α- and β-amyrin (Fig. **13**-7 and **13**-8, respectively), and lupeol (Fig. **13**-9); friedelan-3-one (Fig. **13**-10); 28-hydroxy-friedelan-3-one (Fig. **13**-11), α and β D-glucose [117] (Fig. **13**-12 and **13**-13, respectively).

Indeed, various flavones, including homoorientin (Fig. **13**-14), orientin (Fig. **13**-15), taxifolin deoxyhexose, and isovitexin (Fig. **13**-16), various flavanol derivatives, including (+)-catechin (Fig. **13**-17), (−)-epicatechin (Fig. **13**-18), procyanidin dimers and trimers, and phenolic acids, including protocatechuic (Fig. **13**-19), *P*-hydroxybenzoic (Fig. **13**-20), vanillic (Fig. **13**-21), syringic (Fig. **13**-22), and ferulic acids (Fig. **13**-23) were identified in the juice obtained from the fruits of *E. precatoria* [118].

Nevertheless, in other studies, the compost dehydrodiconiferyl alcohol dibenzoate isolated of extract obtained from *E. precatoria* presented only modest antimalarial activity, displaying CI_{50}= 12 μM against clone 3d7 of *Plasmodium* [41, 119]. The *in vivo* activity in mice infected with *P. berghei* and treated with 100 to 500mg/kg was also investigated, with no promising results [41]. Notwithstanding, from *E. precatoria*, lignan dihydroconiferyl dibenzoate and *P*-hydroxybenzoic acid were also isolated, and the latter substance presented moderate antiplasmodial activity [119].

Hydroalcoholic extracts from *E. precatoria* obtained from leaves and stems were submitted to evaluation of leishmanicidal activity against amastigotes of *L. mexicana*, showing CI_{50}>10 μg/mL. In this same study, the antiplasmodial activity of the extracts was evaluated, with results similar to *Leishmania* [120] (IC_{50}>10 μg/mL). The anti-inflammatory [121] and antioxidant [117] effects of this species have already been evaluated; however, no study was found to evaluate its healing potential. It is noteworthy that antimalarial properties have been attributed to this species, but this activity is quite modest [120], and this suggests that more than a direct antiplasmodial effect, it may act in reducing malaria symptoms and can prevent the worsening of the disease due to its antioxidant potential [122, 123].

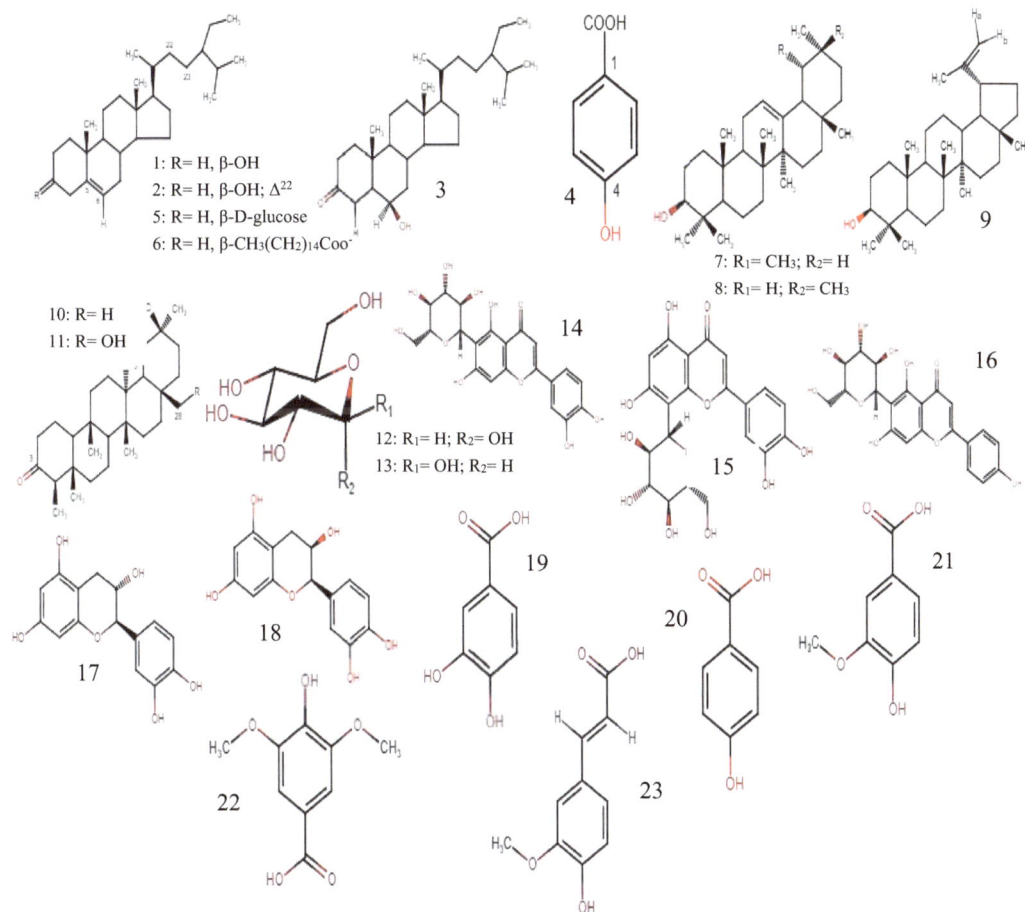

Fig. (13). Chemical structure of compounds isolated from *Euterpe precatoria.*
β-sitosterol (1) and stigmasterol (2), stigmast-4-en-6b-ol-3-one (3), acid *P*-hydroxybenzoic (4), 3b-*O*-b-D-glucopyranosyl sitosterol (5), palmitate β -sitosteril (6), mixture of α- and β -amyrin (7 and 8) and lupeol (9); friedelan-3-one (10); 28-hydroxy-friedelan-3-one (11), α and β D-glucose (12 and 13), homoorientin (14), orientin (15), isovitexin (16); (+)-catechin (17), (−)-epicatechin (18), protocatechuic (19), *P*-hydroxybenzoic (20), vanillic (21), syringic (22), and ferulic acids (23).

Another species often consumed as food by the Amazon population is *Bertholletia excelsa*, the brazilnut tree. Chemical study of extracts from this plant identified the following compounds: gallocatechin (Fig. **14**-1), gallic acid (Fig. **14**-2) and derivatives, protocatechuic acid (Fig. **14**-3), catechin (Fig. **14**-4), protocateualdeyde, protocatechuic acid derivative (Fig. **14**-5), catechin derivative, vanillic acid (Fig. **14**-6) and derivatives, taxifolin (Fig. **14**-7) and derivatives, myricetin-3-o- rhamnoside (Fig. **14**-8), ellagic acid (Fig. **14**-9) and derivatives, and quercetin [124] (Fig. **14**-10).

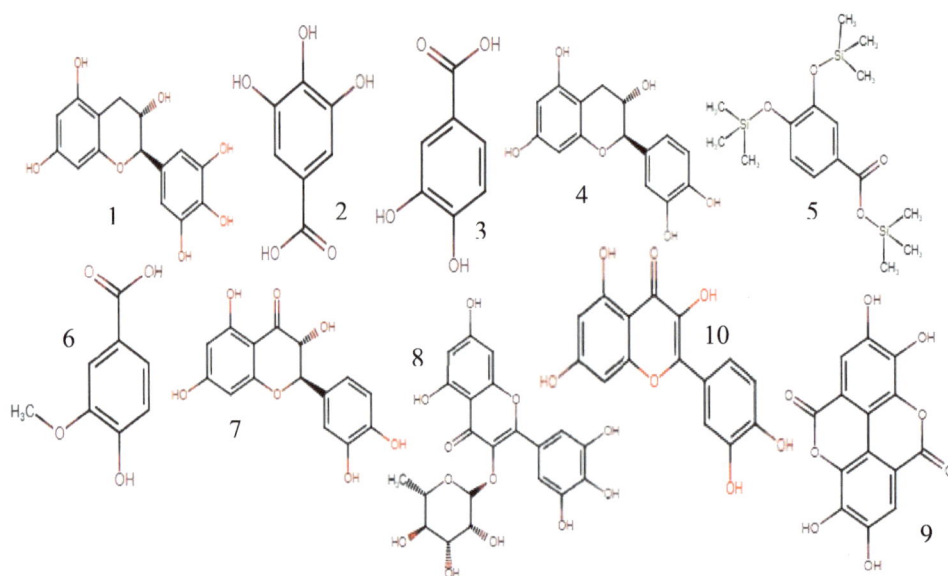

Fig. (14). Chemical structure of compounds isolated from *Bertholletia excelsa*.
gallocatechin (1), gallic acid (2), protocatechuic acid (3), catechin (4), protocatechuic acid derivative (5), vanillic acid (6), taxifolin (7), myricetin-3-o-rhamnoside (8), ellagic acid (9), and quercetin (10).

No studies of antiplasmodial activity of *B. excelsa* were found. Nevertheless, similar to *E. precatoria*, extracts obtained from the kernel and the brown skin that covers the nut of *B. excelsa* were submitted to the evaluation of antioxidant activity. Extracts obtained from the brown skin that covers the nut were more promising as antioxidants, and the activity is related to the higher content of phenolic compounds [124]. The evaluation of the impact of this antioxidant potential on the progression of malaria is important and may lead to the development of new therapeutic uses for this species.

Typically, studies assess the potential for secondary metabolites of plants; however, one study evaluated the antileishmanial potential of proteins from *B. excelsa* and showed that the DR2 fraction presented the strongest toxicity against *L. amazonensis*, causing 100% parasite elimination at 150 µg/mL. DR2 fraction toxic activity included membrane permeabilization, increased endogenous reactive oxygen species (ROS) production, and mitochondrial dysfunction [125].

From the extract of the roots of *Ampelozizyphus amazonicus,* triterpenic saponins-3–O-[β-D-glucopyranosyl(1-2)   α-L-arabine-pyranosyl]-20-O-α-rhamnopy-anosyl-jujubogenine [39] (Fig. **15**-1), and L-ampelozigenin-15α-O-acetyl-3-O α-L-rhamnopyranosyl-(1-2)-β-D glucopyranoside [126] (Fig. **15**-2) were obtained. Additionally, 3-O-[β-D-glucopyranosyl-20-O-α-L rhamnopyranosyl-jujubogenine [127] (Fig. **15**-3) was isolated, as well as terpenoids such as ursolic acid (Fig. **15**-4); betulinic acid (Fig. **15**-5); lupenone (Fig. **15**-6); lupeol (Fig. **15**-7); betulin (Fig. **15**-8); 3β-hydroxylup-20(29)-ene-27,28-dioc acid (Fig. **15**-9); 2α,3β-dihydroxylup-20(29)-ene-27,28- dioic acid (Fig. **15**-10) and 3β,28-dihydroxy-l-p-20(29)-ene-27-oic acid (Fig. **15**-11). Steroids have also been isolated, such as stigmasterol (Fig. **15**-12), sitosterol (Fig. **15**-13), and campesterol [128] (Fig. **15**-14).

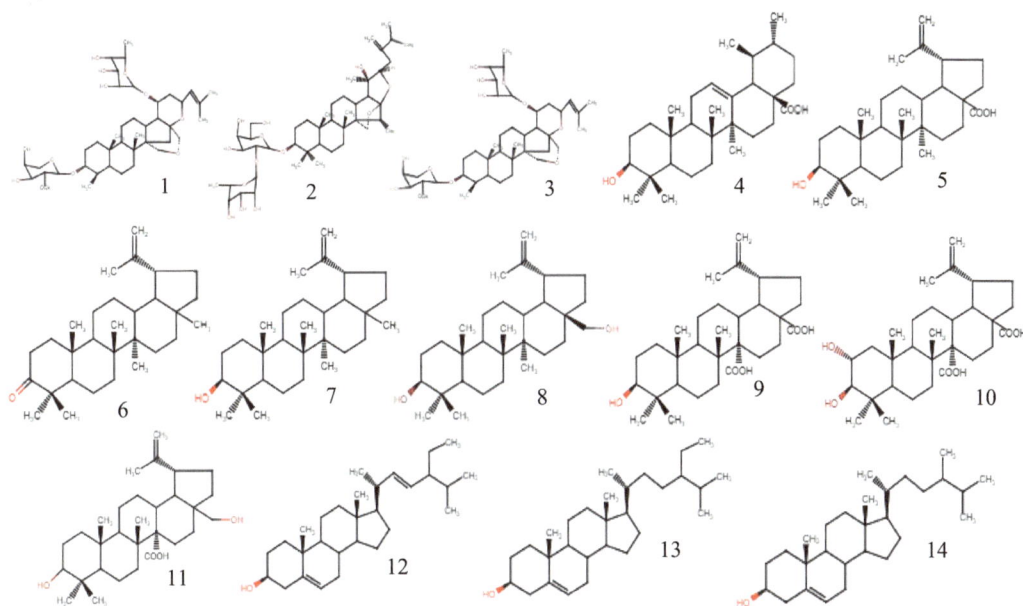

Fig. (15). Chemical structure of compounds isolated from *Ampelozizyphus amazonicus.*
triterpenic saponins: 3–O-[β-D-glucopyranosyl(1-2)α-L-arabine-pyranosyl]-20-O-α -rhamnopyranosy--jujubogenine (1); L-ampelozigenin-15α-O-acetyl-3-O α-L-rhamnopyranosyl-(1-2)-β-D glucopyranoside (2); 3–O-[β-D-glicopyranosyl-20-O-α-L-rhamnopyranosyl-jujubogenine (3), ursolic acid (4), betulinic acid (5), lupenone (6), lupeol (7), betulin (8), 3β-hydroxylup-20(29)-ene-27, 28-dioic acid (9), 2α, 3β-hydroxylu--20(29)-ene-27,28-dioic acid (10), 3β, 28-hdihydroxi-lup-20(29)-ene-27,28-oic acid (11), stigmasterol (12), sitosterol (13), and campesterol (14).

The infusion of roots of *Ampelozizyphus amazonicus* is used in Amazon folk medicine for malaria prevention [129], and the extract of *A. amazonicus* was active against sporozoites of *P. gallinaceum*, as well as during the early stages of the liver cycle [130]. Chloroform and aqueous extracts obtained from *A.*

amazonicus were also tested against *P. berghei* and were also active. Chloroform extract exhibited the highest antiplasmodial activity during the erythrocytic phase of *P. falciparum* and the fractionation of this extract led to the isolation and elucidation of pentacyclic triterpenes, lupeol, botulin, and betulinic acid, which showed high antiplasmodial activity [131].

Extracts from *A. amazonicus* were also tested against promastigote forms of the *Leishmania* species *L. amazonensis, L. braziliensis* and *L. donovani*. Ethanol extract was active against *L. braziliensis* and *L. donovani*, while dichloromethane extract was active only against *L. braziliensis* [132].

As stated earlier, *copaíba* oil (*Copaifera*) is used for wound treatment and as a healing agent. A comprehensive review of the chemical aspects and antileishmanial activity of this oil was conducted by Albuquerque *et al.*, (2017) [133]. Several sesquiterpenes and diterpenes were isolated from this oil (Table **4**; Figs. **16** and **17**, respectively).

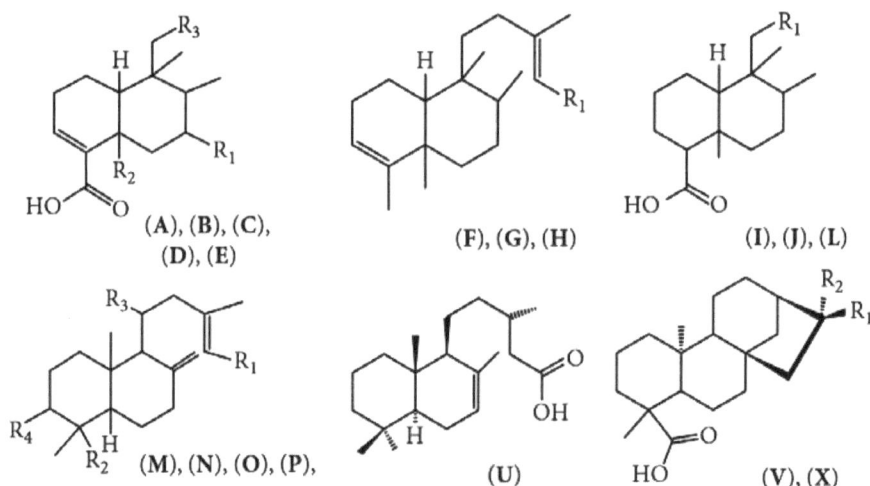

Fig. (16). Diterpenes found in *copaíba* oils [133].
(A) patagonic acid [R_1 = H; R_2= CH_3; R_3 = furanone]; **(B)** hardwickiic acid [R_1 = COOH; R_2 = H; R_3 = furan]; **(C)** 15,16-epoxy-7β-acetoxy-3,13(16),14-clerodatriene-18-oic acid [R_1 = H; R_2 = H; R_3 = furan]; **(D)** 7-hydroxyhardwickiic acid [R_1 = OH; R_2 = CH_3 R_3 = furan]; **(E)** clerodane-15,18-dioic acid [R_1 = H; R_2 = CH_3; R_3 = CH(CH_3)CH_2COOCH_3]; **(F)** 3,13-clerodadiene-15-oic acid [R_1 = COOH]; **(G)** colavenol [R_1 = CH_2OH, *trans* C_1, C_2]; **(H)** cis-colavenol [R_1 = CH_2OH *cis* C_1, C_2]; **(I)** 13-clerodane-15,16-olideo-18oic acid [R_1 = furanone]; **(J)** clerodane-15,18-dioic acid [R_1 = CH_2(CH_3)CH_2CH_2COOH]; **(L)** clorechinic acid [R_1 = furan]; **(M)** copaiferolic acid [R_1 = COOH; R_2 = OH; R_3 = H; R_4 = H]; **(N)** copaiferic acid, [R_1 = COOH; R_2 = CH_3; R_3 = H; R_4 = H]; **(O)** 8(17), 13-labdadiene-15-ol [R_1 = CH_2OH; R_2 = CH_3; R_3 = H; R_4 = H]; **(P)** 11-hydroxycopalic acid [R_1 = COOH; R_2 = CH_3; R_3 = OH; R_4 = H]; **(Q)** ent-3-hydroxy-labd 8(17),13-diene--5-oic acid [R_1 = COOH; R_2 = CH_3; R_3 = H; R_4 = OH]; **(R)** ent-agatic acid [R_1 = COOH R_2 = COOH R_3 = H R_4 = H]; **(S)** copalic acid [R_1 = COOH; R_2 = CH_3; R_3 = H; R_4 = H]; **(T)** 11-acetoxy-copalic acid [R_1 = COOH; R_2 = CH_3; R_3 = CO_2CH_3; R_4 = H]; **(U)** cativic acid; **(V)** ent-16(β)-cauranic-19-oic acid [R_1 = CH_3; R_2 = H]; **(X)** ent-caura-16-ene-19-oic acid [R_1 and R_2 = CH_2].

Table 4. Terpenes present in *Copaifera*

SESQUITERPENES	DITERPENES
	Clerodanes
Alloaromadendrene, ar-curcumene, α-bergamotene, β-bergamotene, ar-curcumene, bicyclogermacrene, β-bisabolene, β-bisabolol, cadalene, cadinene, α-cadinene, δ-cadinene, γ-cadinene, α-cadinol, calamenene, caryophyllene, β-caryophyllene, α-caryophyllenol, cedrol, α-cedrene, cyperene, copaene, α-copaene, β-copaene, γ-elemene, β-farnesene, *trans*-β-farnesene, germacrene B, germacrene D, α-guaiene, β-guaiene, y-guaiene, guaiol, humulene, α-humulene, β-humulene, γ-humulene, ledol, longiciyclene, α-multijugenol, t-muurolol, α-muurolene, γ-muurolene, caryophyllene oxide, α-selinene	3,13-clerodadiene-15,16-olide-18-oic acid 3-clerodene-15,18-dioic acid 13-clerodene-15,16-olide-18-oic acid 3,13-clerodadiene-15-oic acid 3,13-clerodadien-15-ol ent-15,16-epoxy-7β-hydroxy-3,13(16),14-clerodatr-en-18-oic acid ent-(19a)-3,13-clerodadien-15-ol ent-neo-4(18), 13-clerodadien-15-ol clerodene-15,18-dioic acid ent-15,16-epoxy-13(16),14-clerodadien-18-oic acid ent-15,16-epoxy-3,13(16),14-clerodatrien-18-oic acid (+)-7β-Acetoxy-15,16-epoxy-3,13(16),14- clerodatrien-18-oic acid
	Labdanes
	ent-3-hydroxy-labda-8(17),13-dien-15-oic acid ent-8(17),13-labdadien-15,19-dioic acid ent-8(17)-labden-15-oic acid ent-8(17)-labden-15,18-dioic acid ent-15,16-epoxy-8(17),13(16),14-labdatrien-18-oic acid 18-hydroxy-8(17),13-labdadien-15-oic acid 8(17), 13E-labdadien-15-oic acid (13S)-7-labden-15-oic acid 3β-hydroxy-15,16-dinorlabda-8(17)-en-13-one 8(17),13-labdadien-15-ol

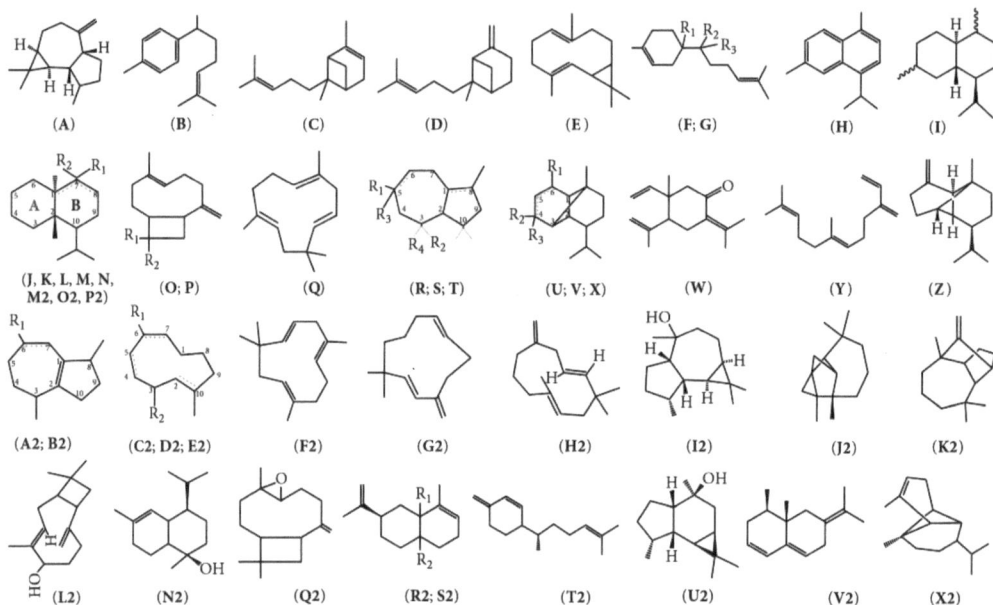

Fig. (17). Sesquiterpenes found in copaiba oils (ALBUQUERQUE *et al.*, 2017) [133].
A) alloaromadendrene; (**B**) ar-curcumene; (**C**) α-bergamotene; (**D**) β-bergamotene; (**E**) bicyclogermacrene; (**F**) β-bisabolene [R_1 = H, R_2 and R_3 = CH_2]; (**G**) β-bisabolol [R_1 = OH, R_2 = CH_3, R_3 = H]; (**H**) cadalene; (**I**) cadinene; (**J**) α-cadinene [R_1 = CH_3; R_2 = not; C_7 = C_8; A = 4- CH_3-hexcycl-3-ene]; (**K**) γ-cadinene [R_1 and R_2 = CH_2; C_7 = C_8; A = 4- CH_3-hexcycl-3ene]; (**L**) $\Box$-cadinene [R_1 = CH_3; R_2 = not; C_1 = C_7; A- CH_3 -hexcycl- 3-ene]; (**M**) α-cadinol [R_1 = H; R_2 = OH; A = 4- CH_3 -hexcycl-3-ene]; (**N**) calamenene [R_1 = H; R_2 = CH_3; A = benzene]; (**O**) caryophyllene [R_1 = CH_3, R_2 = CH_3, *cis*]; (**P**) β-caryophyllene [R_1 = CH_3; R_2 = CH_3, *trans*]; (**Q**) α-caryophyllenol; (**R**) cedrol [R_1 = H; R_2 = CH_3; R_3 = OH; R_4 = CH_3; C_1, C_4 = CH_2]; (**S**) α-cedrene [R_1 = CH_3; R_2 = CH_3; R_3 = not; R_4 = CH_3; C_1, C_4 = CH_2; C_5 = C_6]; (**T**) cyperene [R_1 = H; R_2 = CH_3; R_3 = H; R_4 = C_2, $C_6CH_2(CH_3)_2$]; (**U**) copaene; (**V**) α-copaene; (**X**) β-copaene; (**W**) γ-elemene; (**Y**) β-farnesene; (**Z**) trans-β-farnesene; (**A2**) germacrene B [R_1 = CH_3; R_2 = $C(CH_3)_2$; C_6 = C_7; C_2 = C_{10}]; (**B2**) germacrene D [R_1 = CH_2; C_4 = C_5; C_9 = C_{10}]; (**C2**) α-guaiene [R_1 = $C(CH_2)CH_3$]; (**D2**) β-guaiene [R_1 = $(CH_3)_2$]; (**E2**) γ -guaiene [R_1 = $CH(CH_3)_2$; C_6 = C_7]; (**F2**) humulene; (**G2**) α-humulene; (**H2**) β-humulene; (**I2**) ledol; (**J2**) longicyclene; (**K2**) longifolene; (**L2**) longipinene; (**M2**) α-multijugenol [R_1 = H; R_2 = OH; A = 4-CH_3 - hexcycl-3-ene]; (**N2**) t-muurolol; (**O2**) a-muurolene [R_1 = CH_3; R_2 = not; C_7 = C_8; A = 4-Me-hexcycl-3-ene]; (**P2**): γ-muurolene [R_1 + R_2 = CH_2; A = 4- CH_3-hexcycl-3-ene]; (**Q2**): caryophyllene oxide; (**R2**) α-selinene [R_1 = H; R_2 = CH_3*cis*]; (**S2**) β-selinene [R_1 = H; R_2 = CH_3*trans*]; (**T2**) β-sesquiphellandrene; (**U2**) viridiflorol; (**V2**) β-vetivenene; (**X2**) α-ylangene.

Most studies evaluating antileishmanial activity of *Copaifera* sp. were carried out against strains of *L. amazonensis,* but there is a lack of studies evaluating their activities against intracellular forms of the parasite (amastigote; Table **5**).

The oil of *C. reticulata* displayed activity against the two evolutionary forms of *Leishmania* (promastigote and amastigote) and against two strains: *L. amazonensis* and *L. chagasi* [134, 135] (Table **5**). For the amastigote forms of *L. chagasi,* the oil of *C. reticulata* showed higher activity [135] (CI_{50}= 0.52 µg/mL), while for *L. amazonensis* it showed greater activity against promastigote forms

[134] (CI_{50}= 5.0 µg/mL). The leishmanicidal activity of this species may have been influenced by the chemical composition of the oil. Phytochemical studies carried out from *C. reticulata* demonstrate that this is mainly composed of sesquiterpenes and, among the major constituents, *β*-caryophyllene, trans-*α*-*β*-bergamotene, and bisabolene [136].

Oils obtained from *Copaifera marti, C. cearensis, C. paupera, C. langsdorffii, C. multijuga,* and *C. lucens* have been evaluated primarily against promastigote forms of *L. amazonensis* and all of them showed activity [134] CI_{50}= 10–22 µg/mL), with the only exception being the oil obtained from *Copaifera paupera* [137] (Table **5**). Substances isolated from copaíba oils showed greater activity against amastigote forms of *L. amazonensis*, except for hydroxycopalic acid (Table **5**).

Unlike studies against leishmania, only a few studies show the antimalarial activity of *Copaifera*. In this sense, dichloromethane extract obtained from *Copaifera religiosa* showed promising antiplasmodial activity against strains of *Plasmodium falciparum*, with IC_{50} = 13.4 ± 3.6 µg/mL and 8.5 ± 4.7 µg/mL against chloroquine-sensitive and chloroquine-resistant strains, respectively (Table **5**). However, the methanol extract of *C. religiosa* showed no antiplasmodial activity [138] (CI_{50} = 500.7 ± 16.4 µg/ml and 480.9 ± 34.2 µg/mL for chloroquine-sensitive and resistant strains, respectively).

Additionally, the resin oil of *C. reticulata* containing the sesquiterpenes *β*-caryophyllene (41.7%) and *β*-bisabolene (18.6%) was active against strains of *P. falciparum* [139] (sensitive and resistant to chloroquine; IC_{50} = 1.66 and 2.54 µg/mL, respectively).

Table 5. Antileishmanial and antiplasmodial activity of *Copaifera* and terpenes isolated in this genus.

Species or terpenes isolated	Samples	Antileishmanial activity (IC_{50} µg/mL)		Antiplasmodial activity (IC_{50} µg/mL)		Reference
		Promastigote	Amastigote	Sensitive to chloroquine	Chloroquine resistant	
P. falciparum						
Copaifera religiosa	Dichloromethane	-	-	13.4 ± 3.6 8	8.5 ± 4.7	Lekana-Douki *et al.*, 2011 [138]
Copaifera religiosa	Methanol	-	-	500.7 ± 16.4	480.9 ± 34.2	Lekana-Douki *et al.*, 2011 [138]

(Table 5) cont.....

Species or terpenes isolated	Samples	Antileishmanial activity (IC$_{50}$ µg/mL)		Antiplasmodial activity (IC$_{50}$ µg/mL)		Reference
		Promastigote	**Amastigote**	**Sensitive to chloroquine**	**Chloroquine resistant**	
Copaifera reticulata	Resin Oil	-	-	1.66	2.54	De Souza *et al.*, 2017 [139]
L. chagasi						
Copaifera reticulata	Oil	7.88	0.52		-	Rondon *et al.*, 2012 [135]
L. amazonensis						
Copaifera reticulata	Oil	5.0 ± 0.8	20.0		-	Santos *et al.*, 2008 [134]
Copaifera marti	Oil	14.0 ± 0.9	-		-	Santos *et al.*, 2008 [134]
Copaifera cearensis	Oil	18.0 ± 0.0	-		-	Santos *et al.* 2008 [134]
Copaifera paupera	Oil	11.0 ± 0.4	-		-	Santos *et al.*, 2008 [134]
Copaifera langsdorffii	Oil	20.0 ± 0.8	-		-	Santos *et al.*, 2008 [134]
Copaifera officinalis	Oil	20.0 ± 0.4	-		-	Santos *et al.*, 2008 [134]
Copaifera multijuga	Oil	10.0 ± 0.8	-		-	Santos *et al.*, 2008 [134]
Copaifera lucens	Oil	20.0 ± 0.9	-		-	Santos *et al.*, 2008 [134]
Copaifera paupera	Oil	>100	>100		-	Estevez *et al.*, 2007 [137]
Kaurenoic acid	-	28.0 ± 0.7	3.5 ± 0.5		-	Santos *et al.*, 2013 [140]
Hydroxycopalic acid	-	2.5 ± 0.4	18.0 ± 1.5		-	Santos *et al.*, 2013 [140]
Polyalthic acid	-	35.0 ± 2.0	15.0 ± 1.0		-	Santos *et al.*, 2013 [140]
Pinifolic acid	-	70.0 ± 8.0	4.0 ± 0.4		-	Santos *et al.*, 2013 [140]
Caryophyllene oxide	-	-	2.9		-	Soares *et al.*, 2013 [141]
Sesquiterpenes	-	-	2.3		-	Soares *et al.*, 2013 [141]

(Table 5) cont.....

Species or terpenes isolated	Samples	Antileishmanial activity (IC$_{50}$ µg/mL)		Antiplasmodial activity (IC$_{50}$ µg/mL)		Reference
		Promastigote	Amastigote	Sensitive to chloroquine	Chloroquine resistant	
Amphotericin B	-	0.06 ± 0.0	0.23 ± 0.0	-		Santos *et al.*, 2013 [140]

In the Amazon, the population attributes healing action to the genus *Jatropha*. The alcoholic extract of *J. gossypiifolia* was tested on colonic anastomosis in rats, and only a weak effect was observed in the final stage of healing, but there was a decrease in the inflammation process [142]. In the healing of gastrorrhaphy in rats, a result similar to that of the previous study was seen, with reduction of acute inflammation. Nevertheless, there were no statistical differences compared to the control group [72]. In a third study, similar results were obtained, that is, when evaluating the healing activity of this species in skin wounds in rats, reduced healing potential was obtained [71].

Additionally, some studies attributed antiplasmodial activity to this species. In a study conducted by ONYEGBULE *et al.* (2019) [143] in which the *in vivo* antiplasmodial activity of the ethanol extract from leaves and fractions of *J. gossypiifolia* was evaluated in mice infected with *P. berghei*, it was demonstrated that the fractions of the leaf extract exhibited moderate prophylactic and curative activities, with the ethyl acetate fraction inducing the best antimalarial activity.

MARIZ *et al.* (2010) [144] performed a review intitled "The therapeutic possibilities and toxicological risk of *J. gossypiifolia* L". (Table **6** and Fig. **18**) summarize the compounds already isolated from the species, along with fatty acids that were isolated from the seeds and other metabolites that were found in different parts of the plant and in different types of extracts.

Table 6. Compounds isolated from *Jatropha gossypiifolia*

Class	Vegetable part	Compounds	References
Fatty acids	Seeds	Araquidic acid, Araquidonic acid, Behenic acid, Caprilic acid, Estearic acid, Lignoceric acid, Linoleic acid, Myristic acid, Oleic acid, Palmitic acid, Palmitoleic acid, Ricinoleic acid, Vernolic acid	Ogbobe & Akano, 1993 [145] Prasad *et al.*, 1993 [146] Matos, 2004 [73] Hosamani & Katagi, 2008 [147]

(Table 6) cont.....

Class	Vegetable part	Compounds	References
Alkaloids	Leaves, roots, seeds, and latex	Jatrophine	Morton, 1968 [148] Gupta *et al.*, 1979 [149] Ogbobe & Akano, 1993 [145] Matos, 2004 [73]
Coumarins	Stem, roots, and the whole plant	Cleomiscosin A; 7,8-dihydroxi-6-methoxy-coumarin; Propacin	Das *et al.*, 2003 [150] Das & Kashinatan, 1997 [151]
Diterpenes	Roots, leaves, and the whole plant	Jatropholone A; 2α-OH-Jatrophobon 2β-OH-Jatrophobon 2β-OH-5,6-Isojatropone; Jatrophone; Jatropholone B; Jatrophenone	Adesina, 1982 [152] Taylor *et al.*, 1983 [153] Das & Kashinatan, 1997 [151] Ravindranath *et al.* 2003 [154] Matos, 2004 [73]
Flavonoids	Leaves, roots, stem, seeds, and the whole plant	Apigenin; Ferulic acid; 2,3-bis- (hydroxymethyl)-6,7-methylenedioxy-1-(3'4'-dimethoxyphenyl)-naphthalene; Dihydroprasantaline; Gadaine; Gossypidiene; Gossypifane; Isogadaine, Isovitexin, Vitexin	Subramanian *et al.* 1971 [155] Banerji *et al.* 1984 [156] Das *et al.* 1996[a] [157] Das *et al.*, 1996[b] [158] Das & Das, 1995 [159] Das & Kashinatan, 1997 [151] Das & Anjani, 1999 [160] Matos, 2004 [73]

•

(Table 6) cont.....

Class	Vegetable part	Compounds	References
Lignans	Stem, roots, seeds, and whole plant	Jatrodiene; Jatrophan; Jatrophatrione; Lignan aril naphthalene; 2-Piperonilideno3-veritril-3R-ybutyrolactone; Prasantaline; Tetradecyl ferulate (Tetradecyl (E)- ferulate)	Das & Kashinatham, 1997 [151] Kavitha *et al.*, 1999 [161] Chatterjee *et al.*, 1988 [162] Matos, 2004 [73] Das & Banerji, 1988 [163]

To date, no study has specifically evaluated the effect of extracts from *J. gossypiifolia* on the wound healing process caused by *Leishmania*, but studies of its metabolites (alkaloids and phenolic compounds) have shown that they participate in the protective antioxidant activity of higher organisms, as well as in the inhibition of the enzyme acetylcholinesterase, which causes damage to the membranes of *Leishmania* [164]. Thus, it is possible that the species also presents leishmanicidal properties, as suggested in a previous study by CHAN-BACAB and PEÑA-RODRIGUÉZ (2001) [165].

Another genus that deserves attention is *Musa*, used for the treatment of severe wounds among Amazon peoples. From peeled fruits of *M. paradisiaca*, two acyl steryl glycosides, sitoindoside-III and sitoindoside-IV, and two steryl glycosides, sitosterol gentiobioside and sitosterol myo-inosityl- β-D-glucoside were isolated [166] (Fig. **19**). The tetracyclic triterpene isolated from the flowers of *M. paradisiaca* was determined as (24R)-4α,14α.24-trimethyl-5α-cholesta-8,2-(27)-dien-3β-ol [167], and the leishmanicidal activity of the triterpenes isolated from *M. paradisiaca* and the anacardic acid and synthetic derivatives against *Leishmania infantum chagasi* were also tested. It was identified that, except for cycloeucalone, all other compounds (31-norcyclolauddeone, stigmasterol, β-sitosterol and 24-methylene-cycloartanol) from the fruit peel were active against the promastigote form. On the other hand, against the amastigote forms, all other compounds, including anacardic acid, were active, except for 31-norcyclocondeone [168].

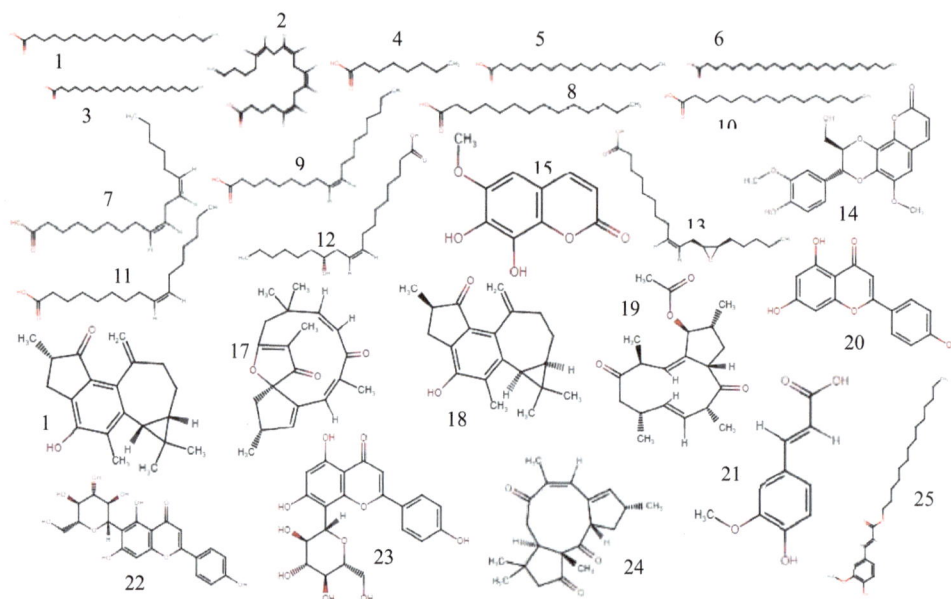

Fig. (18). Chemical structure of compounds isolated from *Jatropha gossypiifolia.*
arachidic acid (1), araquidonic acid (2), behenic acid (3), caprilic acid (4), estearic acid (5), lignoceric acid (6), linoleic acid (7), myristic acid (8), oleic acid (9), palmitic acid, (10), palmitoleic acid (11), ricinoleic acid (12), vernolic acid (13), cleomiscosin a (14), 7,8-dihydroxi-6- methoxy-coumarin (15), jatropholone a (16), jatrophone (17), jatropholone b (18), jatrophenone (19), apigenin (20), ferulic acid (21), isovitexin (22), vitexin (23), jatrophatrione (24), and tetradecyl ferulate (25).

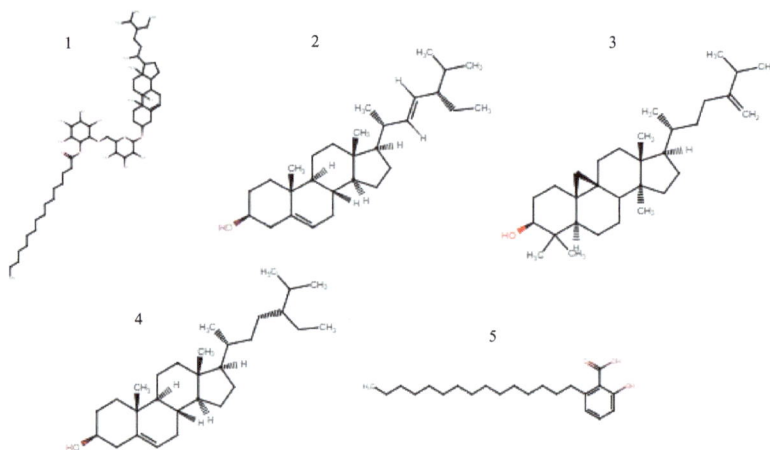

Fig. (19). Chemical structure of compounds isolated from *Musa paradisiaca.*
sitoindoside-IV (1), stigmasterol (2), 24-methylene-cycloartanol (3), β-sitosterol (4), anacardic acid (5).

In summary, some plant species seem to be more promising for malaria, and others for leishmania. Notwithstanding, some species seem to be very promising both for the treatment of malaria and leishmaniasis and these are widely used in folk medicine by Amazonian communities, *i.e.*, species of *Copaifera* and *Aspidosperma nitidum*. The antiparasitic activity of *A. nitidum* is probably related to its alkaloids [169], while the activity of *Copaifera* is related to terpenes (Table 7). It is emphasized that there are other species used by Amazonian communities for the treatment of malaria, including *G. sericeum* and *A. excelsum*, that also produce alkaloids, and these are responsible for the antiparasitic activity [103, 170].

Table 7. Analysis of whether the species is promising for antimalarial or antileishmanial activities and the possible activity marker

Species	Antimalarial	Antileishmanial	Marker	References
Portulaca pilosa	Promising	Promising as a healing agent	NC	Brandão *et al.*, 2020 [169]
Aspidosperma nitidum	Promising	Promising	Tannins, flavonoids, alkaloids, triterpenoids and saponins	Komlaga *et al.*, 2015 [112] Kenechukwu, 2020 [113]
Persea americana	Promising	Extract with moderate activity; Acetogenins were active	5E,12Z,15Z)-2-hydroxy-4-oxohenicosa-5,12,15-triene-1-yl acetate and (2E,5E,12Z,15Z)-1-hydroxyhenicosa-2,5,12-15-triene-4-one	Dharmaratne *et al.*, 2012 [115]
Euterpe precatoria	Modest antiparasitic activity	Modest leishmanicidal activity	β-sitosterol and stigmasterol, stigmast-4-en-6b-ol-3-one, p-hydroxybenzoic acid, 3b-O-b-D- glucopyranosyltestosterol, palmitate β-sitosteryl, mixture of α- and β-amyrin and lupeol; friedelan-3-ona; 28-hydroxy-friedelan-3-ona	Galotta & Boaventura, 2005 [117]
Bertholletia excelsa	NC	Promising	Protein	Fardin *et al.*, 2016 [125]
Ampelozizyphus amazonicus	Promising	Promising	Quercetin 3,3'-dimethyl ether 7-0-α-L-rhamnopyran-syl-(1→6)-β-D-glucopyranose; Quercetin 3,3'-dimethyl ether 7-0-β-D-glucopyranose;	Krettli *et al.*, 2001 [130] do Carmo *et al.*, 2015 [131] Rojas *et al.*, 2009 [132]
Copaifera sp.	Promising	Promising	Rutin, quercetin-3-O-alpha-L rhamnopyranoside, canferol 3-O-alpha-L rhamnopyranoside, quercetin, canferol, abergamotene, α-himachalene, β-selinene, β-caryofilenol, abiotic, daniellic, lambertinnic, labd-7-en-15-oic, isopimaric acids; kaur16-en18-oic, 9.10-Dimethil-1.2 benzanthracene, 3-O-alpha rhamnopyranosil-quercetin, 3-O-alpha rhamnopyranosil-canferol and canferol	Santana *et al.*, 2014 [66] Cavalcante *et al.*, 2017 [68] de Souza *et al.*, 2017 [139]
Jatropha gossypiifolia	Modest antiplasmodial activity	Leishmanicidal effect	Alkaloids (jatrophan glycoside), steroids (β-sitosterol), triterpenoids, saponins and phenolic compounds.	Onyegbule *et al.*, 2019 [143] Martins *et al.*, 2018 [164] Chan-Bacab & Peña-Rodriguéz, 2001 [165]

(Table 7) cont.....

Species	Antimalarial	Antileishmanial	Marker	References
Musa paradisiaca	Modest antiplasmodial activity	Leishmanicidal effect	gentilebioside sitosterol; myo-inosityl-β-D-glucoside sitosterol; (24R)-4α,14α,24-trimethyl-5α-cholesta-8,2-(27)-dien-3β-ol; stigmasterol; β-sitosterol; cycloeucaleone; 31-norcyclolaudeone; 24-methylene-cycloartanol, anacardic acid	Ghosal, 1985 [166] Dutta *et al.*, 1983 [167] Silva *et al.*, 2014 [168] Bagavan *et al.*, 2011 [171]

NC- not yet known.

CONCLUSION

Different species are used by the Amazon population for the treatment of malaria and leishmaniasis. It is observed that species that have alkaloids, such as *G. sericeum, A. nitidum,* and *A. excelsum,* were shown to be promising as antimalarial and antileishmanial agents, and these activities were related to alkaloids. It is emphasized that the antimalarial drug quinine was isolated from a medicinal plant and this reinforces the hypothesis that compounds obtained from these species may be promising for the treatment of both diseases.

In the case of *Portulaca pilosa*, which contains diterpenes, studies have confirmed their antimalarial potential. Moreover, leishmaniasis can be characterized by wounds that are difficult to heal and the use of *P. pilosa* may contribute to the healing process, although it does not seem to have a direct effect against the parasite.

Undoubtedly, copaíba oil seems to be the most promising for the treatment of leishmaniasis since it displays an antiparasitic and healing effect. Another promising species is *Persea americana,* and it seems that acetogenins are the constituents responsible for the leishmanicidal effect. However, it is emphasized that the acetogenin content of these plants, in general, is extremely low and the synthesis is also overly complex. All other species seem to be less promising as antimalarials and leishmanicidal agents.

CONSENT FOR PUBLICATION

Not applicable.

CONFLICT OF INTEREST

The authors declare no conflict of interest, financial or otherwise.

ACKNOWLEDGEMENTS

Authors are grateful to Dr. Pedro Pompei Filizzola Oliva, Dr. Paula Maria Correa de Oliveira, Dr. Pedro Glecio Costa Lima, Dr. Marlia Regina Coelho Ferrreira, and Museu Paraense Emilio Goeldi for the pictures of the vegetal specimens displayed. This work was carried out with the support of CAPES, an entity of the Brazilian Government focused on the training of human resources.

REFERENCES

[1]   Albagli S. Amazônia: fronteira geopolítica da biodiversidade. Parcerias estratégicas 2010; 6(12): 05-19.

[2]   Heck E, Loebens F, Carvalho PD. Amazônia indígena: conquistas e desafios. Estud Av 2005; 19(53): 237-55.
[http://dx.doi.org/10.1590/S0103-40142005000100015]

[3]   Santos FSDD. Tradições populares de uso de plantas medicinais na Amazônia. Hist Cienc Saude Manguinhos 2000; 6: 919-39.
[http://dx.doi.org/10.1590/S0104-59702000000500009]

[4]   Camargo EP. Malária, maleita, paludismo. Cienc Cult 2003; 55(1): 26-9.

[5]   Ferreira AB. Plantas utilizadas no tratamento de malária e males associados por comunidades tradicionais de Xapuri, AC e Pauini, AM. Thesis (PhD) - Universidade Estadual Paulista, Faculdade de Ciências Agronômicas de Botucatu, 2015. Available at: http://hdl.handle.net/11449/126371

[6]   Olson SH, Gangnon R, Silveira GA, Patz JA. Deforestation and malaria in Mâncio Lima County, Brazil. Emerg Infect Dis 2010; 16(7): 1108-15.
[http://dx.doi.org/10.3201/eid1607.091785] [PMID: 20587182]

[7]   Dolabela MF, Oliveira SG, Peres JM, Nascimento JM, Póvoa MM, Oliveira AB. *In vitro* antimalarial activity of six Aspidosperma species from the state of Minas Gerais (Brazil). An Acad Bras Cienc 2012; 84(4): 899-910.
[http://dx.doi.org/10.1590/S0001-37652012000400005] [PMID: 23207699]

[8]   Coutinho JP, Aguiar ACC, dos Santos PA, *et al.* Aspidosperma (Apocynaceae) plant cytotoxicity and activity towards malaria parasites. Part I: Aspidosperma nitidum (Benth) used as a remedy to treat fever and malaria in the Amazon. Mem Inst Oswaldo Cruz 2013; 108(8): 974-82.
[http://dx.doi.org/10.1590/0074-0276130246] [PMID: 24402150]

[9]   Koch I, Rapini A, Simões AO, Kinoshita LS, Spina AP, Castello ACD. The complete reference is: Koch I, Rapini A, Simões AO, Kinoshita LS, Spina AP, Castello ACD. Apocynaceae. In: Lista de Espécies da Flora do Brasil. Jardim Botânico do Rio de Janeiro. Available at: http://floradobrasil.jbrj.gov.br/jabot/floradobrasil/FB48

[10]   Oliveira DRD, Costa ALMA, Leitão GG, Castro NG, Santos JPD, Leitão SG. Estudo etnofarmacognóstico da saracuramirá (*Ampelozizyphus amazonicus* Ducke), uma planta medicinal usada por comunidades quilombolas do Município de Oriximiná-PA, Brasil. Acta Amazon 2011; 41(3): 383-92.
[http://dx.doi.org/10.1590/S0044-59672011000300008]

[11]   Kffuri CW, Lopes MA, Ming LC, Odonne G, Kinupp VF. Plantas antimaláricas usadas por indígenas do Alto Rio Negro no Amazonas, Brasil. J Ethnopharmacol 2016; 178: 188-98.
[http://dx.doi.org/10.1016/j.jep.2015.11.048] [PMID: 26656535]

[12]   Kffuri CW, Ming L, Avila M, Kinupp V, Hidalgo A. Fitonímia Nheengatu de plantas utilizadas no tratamento da malária no alto rio negro–Amazônia brasileira. Ethnoscientia 2020, 5(1). Available at: http://www.ethnoscientia.com/index.php/revista/article/view/27410.22276/ethnoscientia.v5i1.274

[13]   Veiga JB, Scudeller VV. Etnobotânica e medicina popular no tratamento de malária e males associados na comunidade ribeirinha Julião–baixo Rio Negro (Amazônia Central). Rev Bras Plantas Med 2015; 17(4): 737-47.

[http://dx.doi.org/10.1590/1983-084X/14_039]

[14]   Tomchinsky B, Ming LC, Kinupp VF, Hidalgo ADF, Chaves FCM. Ethnobotanical study of antimalarial plants in the middle region of the Negro River, Amazonas, Brazil. Acta Amazon 2017; 47(3): 203-12.
[http://dx.doi.org/10.1590/1809-4392201701191]

[15]   Correa MP. Dicionário de Plantas uteis do Brasil e das Exóticas e Cultivadas. 3 ed., 1–6. Rio de Janeiro: Instituto Brasileiro de Desenvolvimento Florestal 1975.

[16]   Cruz GL, da Silva AC, Eds. Dicionário das plantas úteis do Brasil. Rio de Janeiro: Verlagnichtermittelbar 1979; p. 599.

[17]   Balbach A, Ed. A flora medicinal na medicina doméstica. 17th ed. São Paulo: Edições A Edificação do Lar 1980; p. 921.

[18]   Milliken W. Malaria and antimalarial plants in Roraima, Brazil. Trop Doct 1997; 27(1) (Suppl. 1): 20-5.
[http://dx.doi.org/10.1177/00494755970270S108] [PMID: 9204719]

[19]   Mbeunkui F, Grace MH, Lila MA. Isolation and structural elucidation of indole alkaloids from Geissospermum vellosii by mass spectrometry. J Chromatogr B Analyt Technol Biomed Life Sci 2012; 885-886: 83-9.
[http://dx.doi.org/10.1016/j.jchromb.2011.12.018] [PMID: 22226768]

[20]   Camargo MRM, Amorin RCN, Silva LFR, Carneiro ALB, Vital MJS, Pohlit AM. Chemical composition, ethnopharmacology and biological activity of *Geissospermum allemao* species (Apocynaceae Juss.). Rev Fitos 2013; 8(2): 73-160.

[21]   Borchsenius F, Pedersen HB, Balslev H, Eds. Manual to the Palms of Ecuador. Aarhus: Aarhus University Press 1998; pp. 1-217.

[22]   Vásquez SPF, Mendonça MSD, Noda SDN. Etnobotânica de plantas medicinais em comunidades ribeirinhas do Município de Manacapuru, Amazonas, Brasil. Acta Amazon 2014; 44(4): 457-72.
[http://dx.doi.org/10.1590/1809-4392201400423]

[23]   Ferreira A, Ming LC, Haverroth M, Daly D, Caballero J, Ballesté A. Plants used to treat malaria in the regions of Rio Branco-Acre State and Southern Amazonas State-Brazil. Int J Phytocosm Nat Ingred 2015; 21(1): 2-9.
[http://dx.doi.org/10.15171/ijpni.2015.09]

[24]   Pinheiro KTJDS. Espécies de uso medicinal comercializadas em duas feiras de Manaus-AM. Instituto Federal de Educação, Ciência e Tecnologia do Amazonas, Manaus –AM, 2018. Available at: http://repositorio.ifam.edu.br/jspui/bitstream/4321/264/1/TCC%20FINAL%20KENADY%202018.pdf

[25]   Lima RBD. Flora da reserva Ducke, Amazonas, Brasil: Rhamnaceae. Rodriguésia 2006; 57(2): 247-9.
[http://dx.doi.org/10.1590/2175-7860200657209]

[26]   Krettli AU, Andrade-Neto VF. Search antimalarial drugs in the folk medicine. Ciência Hoje 2004; 35: 70-3.

[27]   Silva JR, Correa GM, Carvalho R, *et al.* Analyses of *Ampelozizyphus amazonicus*, a plant used in folk medicine of the Amazon Region. Pharmacognosy 2009; 5(17): 75-80.

[28]   Santos FSD, Muaze MAF, Eds. Traditions in movement: an ethnohistory of the health and illness in the valleys of the Acre and Purus rivers. 2002.

[29]   Oliveira DR. Ethnobotanical survey of medicinal plants used in the city of Oriximiná (Pará state) with ethnopharmacology focus to the Lippia genus. Master thesis. Universidade Federal do Rio de Janeiro, Rio de Janeiro, 2004, 149 pp (Portuguese).

[30]   Rodrigues E. Plants and animals utilized as medicines in the Jaú National Park (JNP), Brazilian Amazon. Phytother Res 2006; 20(5): 378-91.
[http://dx.doi.org/10.1002/ptr.1866] [PMID: 16619367]

[31]   de Andrade-Neto VF, Pohlit AM, Pinto ACS, *et al.* In vitro inhibition of *Plasmodium falciparum* by substances isolated from Amazonian antimalarial plants. Mem Inst Oswaldo Cruz 2007; 102(3): 359-65.
[http://dx.doi.org/10.1590/S0074-02762007000300016] [PMID: 17568942]

[32]   Altschul SVR, Ed. Drugs and foods from little known plants: notes in Harvard University Herbaria (No 04; QK99, A5). Cambridge: Harvard University Press 1973; p. 366.
[http://dx.doi.org/10.4159/harvard.9780674729209]

[33]   Brandão DLN, Martins MT, Almeida AD, *et al.* Anti-malarial activity and toxicity of *Aspidosperma nitidum* Benth: a plant used in traditional medicine in the Brazilian Amazon. Res Soc Develop 2020; 9: e5059108817.
[http://dx.doi.org/10.33448/rsd-v9i10.8817]

[34]   Milliken W, Albert B. The use of medicinal plants by the Yanomami Indians of Brazil. Econ Bot 1996; 50(1): 10-25.
[http://dx.doi.org/10.1007/BF02862108]

[35]   Scudeller VV, Veiga JD, Araújo-Jorge LD. Etnoconhecimento de plantas de uso medicinal nas comunidades São João do Tupé e Central (Reserva de Desenvolvimento Sustentável do Tupé). In: Santos-Silva EN, Scueller VV, Eds. Biotupé: meio físico, diversidade biológica e sociocultural do Baixo Rio Negro, Amazônia Central. Manaus: UEA Edições 2009; Vol. 2: pp. 185-9.

[36]   Killeen TJ, García E, Beck SG, Eds. Guía de árboles de Bolivia (No C/581984 G8). La Paz: Editorial del Instituto de Ecología 1993; p. 958.

[37]   Le Cointe P, Ed. Amazônia brasileira III Árvores e Plantas úteis (indígenas e aclimadas). 2nd ed. São Paulo: Companhia Editora Nacional Brasiliana 1947; p. 506.

[38]   Frausin G, Hidalgo AdeF, Lima RBS, *et al.* An ethnobotanical study of anti-malarial plants among indigenous people on the upper Negro River in the Brazilian Amazon. J Ethnopharmacol 2015; 174: 238-52.
[http://dx.doi.org/10.1016/j.jep.2015.07.033] [PMID: 26216513]

[39]   Brandão MGL, Grandi TSM, Rocha EMM, Sawyer DR, Krettli AU. Survey of medicinal plants used as antimalarials in the Amazon. J Ethnopharmacol 1992; 36(2): 175-82.
[http://dx.doi.org/10.1016/0378-8741(92)90018-M] [PMID: 1608275]

[40]   Vigneron M, Deparis X, Deharo E, Bourdy G. Antimalarial remedies in French Guiana: a knowledge attitudes and practices study. J Ethnopharmacol 2005; 98(3): 351-60.
[http://dx.doi.org/10.1016/j.jep.2005.01.049] [PMID: 15814272]

[41]   Deharo E, Bourdy G, Quenevo C, Muñoz V, Ruiz G, Sauvain M. A search for natural bioactive compounds in Bolivia through a multidisciplinary approach. Part V. Evaluation of the antimalarial activity of plants used by the Tacana Indians. J Ethnopharmacol 2001; 77(1): 91-8.
[http://dx.doi.org/10.1016/S0378-8741(01)00270-7] [PMID: 11483383]

[42]   Hidalgo ADF. Plantas de uso popular para o tratamento da malária e males associados da área de influência do Rio Solimões e Região de Manaus-AM. PhD Thesis, University of São Paulo State (UNESP), 2003.

[43]   Bertani S, Bourdy G, Landau I, Robinson JC, Esterre P, Deharo E. Evaluation of French Guiana traditional antimalarial remedies. J Ethnopharmacol 2005; 98(1-2): 45-54.
[http://dx.doi.org/10.1016/j.jep.2004.12.020] [PMID: 15849870]

[44]   Balslev H, Grandez C, Zambrana NYP, Møller AL, Hansen SL. Palmas (Arecaceae) útiles en los alrededores de Iquitos, Amazonía Peruana. Rev Peru Biol 2008; 15: 121-32.

[45]   Hajdu Z, Hohmann J. An ethnopharmacological survey of the traditional medicine utilized in the community of Porvenir, Bajo Paraguá Indian Reservation, Bolivia. J Ethnopharmacol 2012; 139(3): 838-57.
[http://dx.doi.org/10.1016/j.jep.2011.12.029] [PMID: 22222280]

[46]   de Oliveira EPB, Peixoto LS, Baldissera M, Andrighetti CR. Uso, diversidade e conhecimento etnobotânico de plantas medicinais utilizadas para o tratamento da malária no município de nova santa helena-MT. FLOVET 2016; 1(8): 89-108.

[47]   Oliveira DR, Krettli AU, Aguiar ACC, *et al.* Ethnopharmacological evaluation of medicinal plants used against malaria by quilombola communities from Oriximiná, Brazil. J Ethnopharmacol 2015; 173: 424-34.
[http://dx.doi.org/10.1016/j.jep.2015.07.035] [PMID: 26231451]

[48]   Cajaiba RL, da Silva WB, de Sousa RDN, de Sousa AS. Levantamento etnobotânico de plantas medicinais comercializadas no município de Uruará, Pará, Brasil. Biotemas 2016; 29(1): 115-31.
[http://dx.doi.org/10.5007/2175-7925.2016v29n1p115]

[49]   Braga R, Ed. Plantas do nordeste, especialmente do Ceará. 2nd ed., Fortaleza: Imprensa Oficial 1960.

[50]   Paes LS, Mendonça MS. Aspectos morfoanatômicos de *Bonamia ferruginea* (Choisy) Hallierf. (Convolvulaceae). Rev Bras Plantas Med 2008; 10: 76-82.

[51]   Coutinho DF, Travassos LMA, do Amaral FMM. Estudo etnobotânico de plantas medicinais utilizadas em comunidades indígenas no Estado do Maranhão-Brasil. Visão Acadêmica 2002; 3(1): 7-12.
[http://dx.doi.org/10.5380/acd.v3i1.493]

[52]   Blair S, Calle BM, Eds. Plantas antimaláricas de Tumaco: costa pacífica colombiana. Mendellín: Editorial Universidad de Antioquia 2005. pp. 347.

[53]   Coelho-Ferreira M. Medicinal knowledge and plant utilization in an Amazonian coastal community of Marudá, Pará State (Brazil). J Ethnopharmacol 2009; 126(1): 159-75.
[http://dx.doi.org/10.1016/j.jep.2009.07.016] [PMID: 19632314]

[54]   Arevalo VG, Ed. Las plantas medicinales y su beneficio en la salud Shipibo-Conibo. Lima: AIDESP 1994; p. 354.

[55]   Delorme RJ, Miolla H, Eds. Pronto socorro do sertão: a cura pelas plantas. Porto Alegre: Escola Superior de Teologia São Lourenço de Brindes 1979; p. 120.

[56]   Neves ES. Introdução ao levantamento da flora medicinal de Rondônia. Porto Velho: Secretaria de Ciência e Tecnologia/Secretaria de Saúde 1980.

[57]   Da Silva FA, Langeloh A, Gonzalez O, Petrovick P. Obtenção e caracterização de extratos de Portulaca pilosa (Amor-crescido) XV Simpósio de Plantas Medicinais do Brasil. Águas de Lindóia: Programa e Resumos 1998; p. 185.

[58]   Souza CCV. Etnobotânica de quintais em três comunidades ribeirinhas na Amazônia Central, Manaus-AM. Master Thesis. Instituto Nacional da Amazônia 2010. Available at: http://localhost:8080/tede/handle/tede/974.

[59]   Paulino-Filho HF, Gottlieb HE, Tomika K, Gottlieb OR, Yoshida M, Lemonica IP. Ampelozizyphus amazonicus Ducke Rhamnaceae I Encontro Regional de Química. São Carlos: Ed. Sociedade Brasileira da Química 1979; p. 63.

[60]   Oliveira FQ, Junqueira RG, Stehmann JR, Brandão MGL. Potencial das plantas medicinais como fonte de novos antimaláricos: espécies indicadas na bibliografia etnomédica brasileira. Rev Bras Plantas Med 2003; 5(2): 23-31.

[61]   Lainson R, Shaw JJ. A brief history of the genus Leishmania (Protozoa: Kinetoplastida) in the Americas with particular reference to Amazonian Brazil 1992. Available at: https://patua.iec.gov.br/handle/iec/2597

[62]   Sacks DL. The structure and function of the surface lipophosphoglycan on different developmental stages of Leishmania promastigotes. Infect Agents Dis 1992; 1(4): 200-6.
[PMID: 1365546]

[63]   Nunes RDO. Prospecção etnofarmacológica de plantas medicinais utilizadas pela população

remanescente de quilombolas de Rolim de Moura do Guaporé, Rondônia, Brasil. 2016. Available at: https://www.locus.ufv.br/handle/123456789/9275

[64]   Alves Barros AS, Oliveira Carvalho H, Dos Santos IVF, *et al.* Study of the non-clinical healing activities of the extract and gel of *Portulaca pilosa* L. in skin wounds in wistar rats: A preliminary study. Biomed Pharmacother 2017; 96: 182-90.
[http://dx.doi.org/10.1016/j.biopha.2017.09.142] [PMID: 28987941]

[65]   Roman ALC, Santos JUMD. A importância das plantas medicinais para a comunidade pesqueira de Algodoal. Bol Mus Para Emílio Goeldi Ciênc Nat 2006; 1(1): 69-80.

[66]   Santana SR, Bianchini-Pontuschka R, Hurtado FB, de Oliveira CA, Melo LPR, dos Santos GJ. Uso medicinal do óleo de copaíba (*Copaifera* sp.) por pessoas da melhor idade no município de Presidente Médici, Rondônia, Brasil. Acta Agron 2014; 63(4): 361-6.
[http://dx.doi.org/10.15446/acag.v63n4.39111]

[67]   Oliveira RLCD, Almeida LFP, Durigan MFB, Veridiana S, Barbosa RI. Conhecimento tradicional e usos de copaíba pela comunidade Makuxi Darora na Savana de Roraima. Gaia Sci 2019; 13(2): 64-72.
[http://dx.doi.org/10.22478/ufpb.1981-1268.2019v13n2.46242]

[68]   Cavalcante JW, Cavalcante V, Bieski I. Conhecimento tradicional e etnofarmacológico da planta medicinal copaiba (*Copaifera langsdorffii* Desf.). Biodiversidade 2017; 16(2): 123-32.

[69]   Aquino JU, Czeczko NG, Malafaia O, *et al.* Avaliação fitoterápica da *Jatropha gossypiifolia* L. na cicatrização de suturas na parede abdominal ventral de ratos. Acta Cir Bras 2006; 21(2): 61-6.
[http://dx.doi.org/10.1590/S0102-86502006000800010] [PMID: 16583056]

[70]   Maia JMA, Czeczko NG, Ribas Filho JM, *et al.* Estudo da cicatrização de suturas na bexiga urinária de ratos com e sem a utilização de extrato bruto de *Jatropha gossypiifolia* L. intraperitoneal. Acta Cir Bras 2006; 21 (Suppl. 2): 23-30.
[http://dx.doi.org/10.1590/S0102-86502006000800005] [PMID: 17117274]

[71]   Santos MFDS, Czeczko NG, Nassif PA, *et al.* Avaliação do uso do extrato bruto de *Jatropha gossypiifolia* L. na cicatrização de feridas cutâneas em ratos. Acta Cir Bras 2006; 21 (Suppl. 3): 2-7.
[http://dx.doi.org/10.1590/S0102-86502006000900002] [PMID: 17293931]

[72]   Vale RJ, Czeczko NG, Aquino JU, *et al.* Estudo comparativo do processo de cicatrização de gastrorafias com e sem o uso do extrato de *Jatropha gossypiifolia* L. (arbusto de dor de barriga) em ratos. Acta Cir Bras 2006; 21: 40-8.
[http://dx.doi.org/10.1590/S0102-86502006000900007]

[73]   Matos FDA, Sousa MP, Craveiro AA, Matos MEO, Eds. Constituintes químicos ativos e propriedades biológicas de plantas medicinais brasileiras. Fortaleza: Editora UFC 2004; p. 448.

[74]   Camargo MTLA, Ed. Plantas medicinais e de rituais afro-brasileiros II: estudo etnofarmacobotânico. São Paulo: Ícone 1998; p. 232.

[75]   Lans C, Harper T, Georges K, Bridgewater E. Medicinal and ethnoveterinary remedies of hunters in Trinidad. BMC Complement Altern Med 2001; 1(1): 10.
[http://dx.doi.org/10.1186/1472-6882-1-10] [PMID: 11737880]

[76]   Lorenzi H, Matos FJ, Eds. Plantas medicinais no Brasil: nativas e exóticas. Nova Odessa: Plantarum 2002, pp. 512.

[77]   Sievers AF, Archer AW, Moore RH, McGovran ER. Insecticidal tests of plants from tropical America. J Econ Entomol 1949; 42(3): 549-51.
[http://dx.doi.org/10.1093/jee/42.3.549]

[78]   Rodrigues AP, Andrade LHC. Levantamento etnobotânico das plantas medicinais utilizadas pela comunidade de Inhamã, Pernambuco, Nordeste do Brasil. Rev Bras Plantas Med 2014; 16(3): 721-30.
[http://dx.doi.org/10.1590/1983-084x/08_159]

[79]   Silva AG, Lima RA, de Souza ACR. Uso, conservação e diversidade de plantas aromáticas,

condimentares e medicinais para fins medicinais na comunidade Vila Princesa, Porto Velho-RO/Use, storage and diversity of aromatic herbs, spices and medicinal uses for medical purposes in Vila Princesa, Porto Velho-RO. Rev Pesq Cria 2011; 10(2): 21-35.

[80]   Sarquis RSFR, Sarquis ÍR, Sarquis IR, *et al.* The use of medicinal plants in the riverside community of the Mazagão River in the Brazilian Amazon, Amapá, Brazil: ethnobotanical and ethnopharmacological studies. Evidence-Based Complem Alter Med 2019; Article ID 6087509 https://doi.org/10.1155/2019/6087509.

[81]   Leão RBA, Ferreira MRC, Jardim MAG. Levantamento de plantas de uso terapêutico no município de Santa Bárbara do Pará, Estado do Pará, Brasil. Rev Bras Farm 2007; 88(1): 21-5.

[82]   Mors BW, Rizzini TC, Pereira AN, Eds. Medicinal Plants of Brazil EUA. Reference Publications, Inc. 2000; p. 289.

[83]   Revilla J, Ed. Plantas úteis da bacia amazônica (No 581 R454p). Manaus: Instituto Nacional de Pesquisas da Amazónia 2002.

[84]   Alves AS, Pinheiro EDSRP, de Oliveira Júnior A, Pena F, Udhe M. As dez plantas medicinais mais indicadas pelos curadores tradicionais no estado do Amapá. Rev Ciênc Agrovet 2006; 5(4): 42-52.

[85]   Barata LES, Alencar AAJ, Tascone M, Tamashiro J. Plantas Medicinais Brasileiras. II. *Portulaca pilosa* L. (Amor-crescido). Rev Fitos 2009; 4(1): 126-8.

[86]   Oak G, Kurve P, Kurve S, Pejaver M. Ethno-botanical studies of edible plants used by tribal women of Thane District. J Med Plant Stud 2015; 3(2): 90-4.

[87]   Botsaris AS. Plants used traditionally to treat malaria in Brazil: the archives of Flora Medicinal. J Ethnobiol Ethnomed 2007; 3(18): 18.
       [http://dx.doi.org/10.1186/1746-4269-3-18] [PMID: 17472740]

[88]   Ohsaki A, Kasetani Y. Asaka, Y, Shibata K, Tokoroyama T, Kubota T. A diterpenoid from *Portulaca pilosa*. Phytochemistry 1995; 40(1): 205-7.
       [http://dx.doi.org/10.1016/0031-9422(95)00228-Y] [PMID: 7786488]

[89]   Ohsaki A, Kasetani Y, Asaka Y, Shibata K, Tokoroyama T, Kubota T. Clerodane diterpenoids from the roots of *Portulaca pilosa*. Phytochemistry 1991; 30(12): 4075-7.
       [http://dx.doi.org/10.1016/0031-9422(91)83470-6]

[90]   Veiga ASS. Atividade antileishmania de plantas Amazônicas. Master Thesis, Instituto de Ciências da Saúde, Universidade Federal do Pará, Belém, 2013.

[91]   Verpoorte R, van Beek TA, Thomassen PHAM, Aandewiel J, Baerheim Svendsen A. Screening of antimicrobial activity of some plants belonging to the Apocynaceae and Loganiaceae. J Ethnopharmacol 1983; 8(3): 287-302.
       [http://dx.doi.org/10.1016/0378-8741(83)90066-1] [PMID: 6645578]

[92]   Marques MDFDS. Contribuição ao estudo químico do gênero Aspidosperma: Aspidosperma ramiflorum Muell. Arg. Master Thesis, Universidade Estadual de Campinas, 1988.

[93]   Nunes DS, Koike L, Taveira JJ, Reis FDAM. Indole alkaloids from *Aspidosperma pruinosum*. Phytochemistry 1992; 31: 2507-11.
       [http://dx.doi.org/10.1016/0031-9422(92)83311-L]

[94]   Benoin PR, Burnell RH, Medina JD. Alkaloids from *Aspidosperma excelsum* Benth. Can J Chem 1967; 45(7): 725-30.
       [http://dx.doi.org/10.1139/v67-118]

[95]   Arndt RR, Brown SH, Ling NC, *et al.* Alkaloid studies—LVIII: The alkaloids of six Aspidosperma species. Phytochemistry 1967; 6(12): 1653-8.
       [http://dx.doi.org/10.1016/S0031-9422(00)82898-8]

[96]   Marques MFS, Kato L, Leitão Filho HF, Reis FAM. Indole alkaloids from *Aspidosperma ramiflorum*. Phytochemistry 1996; 41(3): 963-7.

[http://dx.doi.org/10.1016/0031-9422(95)00660-5]

[97] Pereira MDM, Jácome RLRP, Alcântara ADC, Alves RB, Raslan DS. Alcalóides indólicos isolados de espécies do gênero Aspidosperma (Apocynaceae). Quim Nova 2007; 30(4): 970-83. [http://dx.doi.org/10.1590/S0100-40422007000400037]

[98] Nascimento PC, Araújo RM, Silveira ER. Aplicação da CLAE na análise fitoquímica de Aspidosperma nitidum Águas de Lindóia: Reunião Anual da Sociedade Brasileira de Química. São Paulo: Resumos 2006.

[99] Martins MT. Estudo farmacognóstico, fitoquímico e atividades biológicas de Aspidosperma nitidum Benth. ExMull. Arg. Master Thesis, Instituto Ciências da Saúde, Universidade Federal do Pará, Belém-Brazil, 2012.

[100] Veiga ASS. Atividade antileishmania de plantas Amazônicas. Master Thesis, Instituto Ciências da Saúde, Universidade Federal do Pará, Belém-Brazil, 2013. 2013.

[101] Saito ML, Fazolin M, Maia AH N, Horiuchi EYO. Avaliação de atividades biológicas em plantas da região amazônica para controle de insetos. Embrapa Meio Ambiente. Boletim de Pesquisa e Desenvolvimento 2006. p. 18.

[102] Manske RHF, Harrison WA, Eds. The alkaloids of *Geissospermum* species. The Alkaloids: Chemistry and Physiology. Academic Press 1965; Vol. 8: pp. 679-91.

[103] Steele JC, Veitch NC, Kite GC, Simmonds MS, Warhurst DC. Indole and β-carboline alkaloids from *Geissospermum sericeum*. J Nat Prod 2002; 65(1): 85-8. [http://dx.doi.org/10.1021/np0101705] [PMID: 11809075]

[104] Carvalho LH, Brandão MGL, Santos-Filho D, Lopes JLC, Krettli AU. Antimalarial activity of crude extracts from Brazilian plants studied *in vivo* in *Plasmodium berghei*-infected mice and *in vitro* against *Plasmodium falciparum* in culture. Braz J Med Biol Res 1991; 24(11): 1113-23. [PMID: 1823001]

[105] da Silva E Silva JV, Cordovil Brigido HP, Oliveira de Albuquerque KC, *et al.* Flavopereirine - an alkaloid derived from *Geissospermum vellosii* - presents leishmanicidal activity *in vitro*. Molecules 2019; 24(4): 785. [http://dx.doi.org/10.3390/molecules24040785] [PMID: 30795632]

[106] Reina M, Ruiz-Mesia W, López-Rodríguez M, Ruiz-Mesia L, González-Coloma A, Martínez-Díaz R. Indole alkaloids from *Geissospermum reticulatum*. J Nat Prod 2012; 75(5): 928-34. [http://dx.doi.org/10.1021/np300067m] [PMID: 22551062]

[107] Passemar C, Saléry M, Soh PN, *et al.* Indole and aminoimidazole moieties appear as key structural units in antiplasmodial molecules. Phytomedicine 2011; 18(13): 1118-25. [http://dx.doi.org/10.1016/j.phymed.2011.03.010] [PMID: 21612900]

[108] do Nascimento MS, Pina NDPV, da Silva ASB, *et al. In vitro* antiplasmodial activity and identification, using tandem LC-MS, of alkaloids from *Aspidosperma excelsum*, a plant used to treat malaria in Amazonia. J Ethnopharmacol 2019; 228: 99-109. [http://dx.doi.org/10.1016/j.jep.2018.09.012] [PMID: 30201230]

[109] Mitaine-Offer A-C, Sauvain M, Valentin A, Callapa J, Mallié M, Zèches-Hanrot M. Antiplasmodial activity of aspidosperma indole alkaloids. Phytomedicine 2002; 9(2): 142-5. [http://dx.doi.org/10.1078/0944-7113-00094] [PMID: 11995947]

[110] Saxton JE. Recent progress in the chemistry of the monoterpenoid indole alkaloids. Nat Prod Rep 1996; 13(4): 327. [http://dx.doi.org/10.1039/np9961300327] [PMID: 7666980]

[111] Park S, Nam YH, Rodriguez I, *et al.* Constituintes químicos das folhas de *Persea americana* (abacate) e seus efeitos protetores contra o dano às células ciliadas induzido pela neomicina. Rev Bras Farmacogn 2019; 29(6): 739-43. [http://dx.doi.org/10.1016/j.bjp.2019.08.004]

[112] Komlaga G, Cojean S, Beniddir MA, Loiseau PM. The antimalarial potential of three Ghanaian medicinal plants. Herbal Med 2015; 1(4): 1-7.
[http://dx.doi.org/10.21767/2472-0151.10004]

[113] Kenechukwu OC. Evaluation of *in vivo* anti-malarial activity of methanolic leaf extract of *Persea Americana* against *Plasmodium berghei*-infected mice. J Scient Res 2020; 5(1): 44-52.

[114] Falodun A, Erharuyi O, Imieje V, *et al. In vitro* evaluation of aliphatic fatty alcohol metabolites of *Perseaamericana* seed as potential antimalarial and antimicrobial agents. Niger J Biotechnol 2014; 27: 1-7.
[PMID: 28042193]

[115] Dharmaratne HRW, Tekwani B, Jacob MR, Nanayakkara NPD. Antimicrobial and antileishmanial active acetogenins from avocado (*Persea americana*) fruits. Planta Med 2012; 78(05): 34.
[http://dx.doi.org/10.1055/s-0032-1307542]

[116] Ogungbe IV, Setzer WN. The potential of secondary metabolites from plants as drugs or leads against protozoan neglected diseases—Part III: In-silico molecular docking investigations. Molecules 2016; 21(10): 1389.
[http://dx.doi.org/10.3390/molecules21101389] [PMID: 27775577]

[117] Galotta ALQDA, Boaventura MAD. Constituintes químicos da raiz e do talo da folha do açaí (*Euterpe precatoria* Mart., Arecaceae). Quim Nova 2005; 28(4): 610-3.
[http://dx.doi.org/10.1590/S0100-40422005000400011]

[118] Pacheco-Palencia LA, Duncan CE, Talcott ST. Composição fitoquímica e estabilidade térmica de duas espécies comerciais de açaí, *Euterpe oleracea* e *Euterpe precatoria.* Química de Alimentos 2009; 115(4): 1199-205.

[119] Jensen JF, Kvist LP, Christensen SB. An antiplasmodial lignan from *Euterpe precatoria.* J Nat Prod 2002; 65(12): 1915-7.
[http://dx.doi.org/10.1021/np020264u] [PMID: 12502338]

[120] Calderon AI, Romero LI, Ortega-Barria E, Brun R, Correa AMD, Gupta MP. Evaluation of larvicidal and *in vitro* antiparasitic activities of plants in a biodiversity plot in the Altos de Campana National Park, Panama. Pharm Biol 2006; 44(7): 487-98.
[http://dx.doi.org/10.1080/13880200600878361]

[121] Kang J, Thakali KM, Xie C, *et al.* Bioactivities of açaí (*Euterpe precatoria* Mart.) fruit pulp, superior antioxidant and anti-inflammatory properties to *Euterpe oleracea* Mart. Food Chem 2012; 133(3): 671-7.
[http://dx.doi.org/10.1016/j.foodchem.2012.01.048]

[122] Percário S, Moreira DR, Gomes BAQ, *et al.* Oxidative stress in malaria. Int J Mol Sci 2012; 13(12): 16346-72.
[http://dx.doi.org/10.3390/ijms131216346] [PMID: 23208374]

[123] Quadros Gomes BA, da Silva LF, Quadros Gomes AR, *et al.* N-acetyl cysteine and mushroom *Agaricus sylvaticus* supplementation decreased parasitaemia and pulmonary oxidative stress in a mice model of malaria. Malar J 2015; 14: 202.
[http://dx.doi.org/10.1186/s12936-015-0717-0] [PMID: 25971771]

[124] John JA, Shahidi F. Phenolic compounds and antioxidant activity of Brazil nut (*Bertholletia excelsa*). J Funct Foods 2010; 2(3): 196-209.
[http://dx.doi.org/10.1016/j.jff.2010.04.008]

[125] Fardin JM, Carvalho LP, do Nascimento VV, Melo EJT, Gomes VM, Machado OLT. Biochemical purification of proteins from *Bertholletia excelsa* seeds and their antileishmanial action *in vitro.* World J Pharm Res 2016; 5(7): 233-300.

[126] Brandao MG, Lacaille-Dubois MA, Teixeira MA, Wagner H. A dammarane-type saponin from the roots of *Ampelozizyphus amazonicus.* Phytochemistry 1993; 34(4): 1123-7.

[http://dx.doi.org/10.1016/S0031-9422(00)90728-3] [PMID: 7764239]

[127] Diniz LRL. Efeito das saponinas triterpênicas isoladas de raízes da Ampelozizyphus amazonicus Ducke sobre a função renal. Thesis, Departamento de Fisiologia e Biofísica, Instituto de Ciências Biológicas, Universidade Federal de Minas, 2006.

[128] Amaral AC, Ferreira JL, de Moura D, *et al.* Estudos atualizados sobre *Ampelozizyphus amazonicus*, planta medicinal utilizada na Região Amazônica. Pharmacog Ver 2008; 2(4): 308.

[129] Brandão MGL. Estudo Químico da Ampelozizyphus amazonicus Ducke, planta utilizada na Amazônia como preventivo da malária. PhD Thesis, Universidade Federal de Minas Gerais, Belo Horizonte, 1991.

[130] Krettli AU, Andrade-Neto VF, Brandão MDGL, Ferrari W. A busca de novos antimaláricos a partir de plantas utilizadas no tratamento de febre e malária ou plantas selecionadas aleatoriamente: uma revisão. Mem Inst Oswaldo Cruz 2001; 96(8): 1033-42.
[http://dx.doi.org/10.1590/S0074-02762001000800002] [PMID: 11784919]

[131] do Carmo DF, Amaral ACF, Machado M, *et al.* Avaliação da atividade antiplasmódica de extratos e constituintes de *Ampelozizyphus amazonicus*. Pharmacognosy 2015; 11 (Suppl. 2): S244.
[PMID: 26664012]

[132] Rojas U, Satalaya A, Johann R, *et al.* Actividad leishmanicida de plantas medicinales de la Amazonia peruana. Rev Boliv Quím 2009; 26(2): 43-8.

[133] Albuquerque KCOD, Veiga ADSSD, Brigido HPC, *et al.* Brazilian Amazon traditional medicine and the treatment of difficult to heal leishmaniasis wounds with Copaifera. Evid Bas Complem Alter Med 2017; Article ID 8350320, Available at: https://doi.org/10.1155/2017/8350320.

[134] Santos AO, Ueda-Nakamura T, Dias Filho BP, Veiga Junior VF, Pinto AC, Nakamura CV. Effect of Brazilian copaiba oils on *Leishmania amazonensis*. J Ethnopharmacol 2008; 120(2): 204-8.
[http://dx.doi.org/10.1016/j.jep.2008.08.007] [PMID: 18775772]

[135] Rondon FCM, Bevilaqua CML, Accioly MP, *et al.* In vitro efficacy of *Coriandrum sativum*, Lippia sidoides and *Copaifera reticulata* against *Leishmania chagasi*. Rev Bras Parasitol Vet 2012; 21(3): 185-91.
[http://dx.doi.org/10.1590/S1984-29612012000300002] [PMID: 23070424]

[136] Herrero-Jáuregui C, Casado MA, das Graças Bichara Zoghbi M, Célia Martins-da-Silva R. Chemical variability of *Copaifera reticulata* Ducke oleoresin. Chem Biodivers 2011; 8(4): 674-85.
[http://dx.doi.org/10.1002/cbdv.201000258] [PMID: 21480513]

[137] Estevez Y, Castillo D, Pisango MT, *et al.* Evaluation of the leishmanicidal activity of plants used by Peruvian Chayahuita ethnic group. J Ethnopharmacol 2007; 114(2): 254-9.
[http://dx.doi.org/10.1016/j.jep.2007.08.007] [PMID: 17889471]

[138] Lekana-Douki JB, Oyegue Liabagui SL, Bongui JB, Zatra R, Lebibi J, Toure-Ndouo FS. *In vitro* antiplasmodial activity of crude extracts of *Tetrapleura tetraptera* and *Copaifera religiosa*. BMC Res Notes 2011; 4: 506.
[http://dx.doi.org/10.1186/1756-0500-4-506] [PMID: 22112366]

[139] de Souza GA, da Silva NC, de Souza J, *et al.* *In vitro* and *in vivo* antimalarial potential of oleoresin obtained from *Copaifera reticulata* Ducke (Fabaceae) in the Brazilian Amazon rainforest. Phytomedicine 2017; 24: 111-8.
[http://dx.doi.org/10.1016/j.phymed.2016.11.021] [PMID: 28160850]

[140] Santos AOD, Izumi E, Ueda-Nakamura T, Dias-Filho BP, Veiga-Júnior VFD, Nakamura CV. Antileishmanial activity of diterpene acids in copaiba oil. Mem Inst Oswaldo Cruz 2013; 108(1): 59-64.
[http://dx.doi.org/10.1590/S0074-02762013000100010] [PMID: 23440116]

[141] Soares DC, Portella NA, Ramos MFDS, Siani AC, Saraiva EM. Trans-β-Caryophyllene: An effective antileishmanial compound found in commercial copaiba oil (Copaifera spp). Evid Based Compl Alter

Med 2013; p. 761323.

[142] Servin SCN, Torres OJM, Matias JEF, *et al.* Ação do extrato de *Jatropha gossypiifolia* L. (pião roxo) na cicatrização de anastomose colônica: estudo experimental em ratos. Acta Cir Bras 2006; 21 (Suppl. 3): 89-96.
[http://dx.doi.org/10.1590/S0102-86502006000900012] [PMID: 17293941]

[143] Onyegbule FA, Bruce SO, Onyekwe ON, Onyealisi OL, Okoye PC. Evaluation of the *in vivo* antiplasmodial activity of ethanol leaf extract and fractions of *Jatropha gossypiifolia* in *Plasmodium berghei* infected mice. J Med Plants Res 2019; 13(11): 269-79.
[http://dx.doi.org/10.5897/JMPR2019.6766]

[144] Mariz SR, Borges ACR, Melo-Diniz MFF, Medeiros IA. Possibilidades terapêuticas e risco toxicológico de *Jatropha gossypiifolia* L.: uma revisão narrativa. Rev Bras Plantas Med 2010; 12(3): 346-57.
[http://dx.doi.org/10.1590/S1516-05722010000300013]

[145] Ogbobe O, Akano V. The physico-chemical properties of the seed and seed oil of Jatropha gossipifolia. Plant Foods Hum Nutr 1993; 43(3): 197-200.
[http://dx.doi.org/10.1007/BF01886220] [PMID: 8506234]

[146] Prasad YR, Alankararao GSJG, Baby P. Constituents of the seeds of *Jatropha gossypiifolia.* Fitoterapia 1993; 64(4): 376.

[147] Hosamani KM, Katagi KS. Characterization and structure elucidation of 12-hydroxyoctadec-ci--9-enoic acid in Jatropha gossipifolia and Hevea brasiliensis seed oils: a rich source of hydroxy fatty acid. Chem Phys Lipids 2008; 152(1): 9-12.
[http://dx.doi.org/10.1016/j.chemphyslip.2007.11.003] [PMID: 18060875]

[148] Morton JF. A survey of medicinal plants of Curacao. Econ Bot 1968; 22(1): 87-102.
[http://dx.doi.org/10.1007/BF02897749]

[149] Gupta MP, Arias TD, Correa M, Lamba SS. Ethnopharmacognostic observations on Panamanian medicinal plants. Part I. Q J Crude Drug Res 1979; 17(3-4): 115-30.
[http://dx.doi.org/10.3109/13880207909065163]

[150] Das B, Kashinatham A, Venkataiah B, Srinivas KVNS, Mahender G, Reddy MR. Cleomiscosin A, a coumarino-lignoid from *Jatropha gossypiifolia.* Biochem Syst Ecol 2003; 31(10): 1189-91.
[http://dx.doi.org/10.1016/S0305-1978(03)00067-X]

[151] Das B, Kashinatham A. Studies on phytochemicals: Part XVII – Phenolics from the roots of *Jatropha gossypiifolia.* Indian J Chem 1997; 36: 1077-8.

[152] Adesina SK. Studies on some plants used as anticonvulsant in Amerindian and African traditional medicine. Fitoterapia 1982; 53: 147-62.

[153] Taylor MD, Smith AB III, Furst GT, *et al.* Plant anticancer agents. 28. New antileukemic jatrophone derivatives from *Jatropha gossypiifolia*: structural and stereochemical assignment through nuclear magnetic resonance spectroscopy. J Am Chem Soc 1983; 105(10): 3177-83.
[http://dx.doi.org/10.1021/ja00348a036]

[154] Ravindranath N, Venkataiah B, Ramesh C, Jayaprakash P, Das B. Jatrophenone, a novel macrocyclic bioactive diterpene from *Jatropha gossypifolia.* Chem Pharm Bull (Tokyo) 2003; 51(7): 870-1.
[http://dx.doi.org/10.1248/cpb.51.870] [PMID: 12843600]

[155] Subramanian SS, Nagarajan S, Sulochana N. Flavonoids of the leaves of *Jatropha gossypiifolia.* Phytochemistry 1971; 10(7): 1690-0.
[http://dx.doi.org/10.1016/0031-9422(71)85055-0]

[156] Banerji J, Das B, Chatterjee A, Shoolery JN. Gadain, a lignan from *Jatropha gossypiifolia.* Phytochemistry 1984; 23(10): 2323-7.
[http://dx.doi.org/10.1016/S0031-9422(00)80544-0]

[157]   Das B, Rao SP, Srinivas KVNS, Das R. Jatrodien, a lignan from stems of *Jatropha gossypiifolia.* Phytochemistry 1996; 41(3): 985-7. a
[http://dx.doi.org/10.1016/0031-9422(95)00729-6]

[158]   Das B, Rao SP, Srinivas KV. Isolation of isogadain from *Jatropha gossypifolia.* Planta Med 1996; 62(1): 90. b
[http://dx.doi.org/10.1055/s-2006-957818] [PMID: 17252424]

[159]   Das B, Das R. Gossypifan, a lignan from *Jatropha gossypiifolia.* Phytochemistry 1995; 40(3): 931-2.
[http://dx.doi.org/10.1016/0031-9422(95)00400-2]

[160]   Das B, Anjani G. Gossypidien, a lignan from stems of *Jatropha gossypiifolia.* Phytochemistry 1999; 51(1): 115-7.
[http://dx.doi.org/10.1016/S0031-9422(98)00727-4]

[161]   Kavitha J, Rajasekhar D, Subbaraju GV. Synthesis of tetradecyl (E)-ferulate, a metabolite of Jatropha gossypifolia. J Asian Nat Prod Res 1999; 2(1): 51-4.
[http://dx.doi.org/10.1080/10286029908039891] [PMID: 11261206]

[162]   Chatterjee A, Das B, Chakrabarti R, *et al.* 1988.

[163]   Das B, Banerji J. Aryl nafthalene lignan de *Jatropha gossyplifolia.* Phytochemistry 1988; 27(11): 684-3686.
[http://dx.doi.org/10.1016/0031-9422(88)80799-4]

[164]   Martins GV, Alves DR, Viera-Araújo FM, Rondon F. Estudo químico e avaliação das atividades antioxidante, anti-acetilcolinesterase e anti-leishmanial de extratos de *Jatropha gossyplifolia* L(pião roxo) Rev Virt Quím 2018; 10(1)

[165]   Chan-Bacab MJ, Peña-Rodríguez LM. Plant natural products with leishmanicidal activity. Nat Prod Rep 2001; 18(6): 674-88.
[http://dx.doi.org/10.1039/b100455g] [PMID: 11820764]

[166]   Ghosal S. Steryl glicosídeos e glicosides de ciclico de *Musa paradisiaca.* Fitoquímica 1985; 24(8): 1807-10.

[167]   Dutta PK, Das AK, Banerji N. A tetracyclic triterpenoid from *Musa paradisiaca.* Phytochemistry 1983; 22(11): 2563-4.
[http://dx.doi.org/10.1016/0031-9422(83)80165-4]

[168]   Silva AAS, Morais SM, Falcão MJC, *et al.* Activity of cycloartane-type triterpenes and sterols isolated from *Musa paradisiaca* fruit peel against *Leishmania infantum chagasi.* Phytomedicine 2014; 21(11): 1419-23.
[http://dx.doi.org/10.1016/j.phymed.2014.05.005] [PMID: 24916706]

[169]   Brandão DLN, Vale VV, da Veiga ADSS, *et al.* Importância do amor-crescido (*Portulaca pilosa* L.) para a medicina tradicional amazônica: uma revisão bibliográfica. Rev Eletr Acervo Saúde 2020; 12(3): e2371-1.
[http://dx.doi.org/10.25248/reas.e2371.2020]

[170]   Correa-Barbosa J, Silva MCM, Percario P, Brasil DSB, Dolabela MF, Vale VV. *Aspidosperma excelsum* and its pharmacological potential: in silico studies of pharmacokinetic prediction, toxicological and biological activity. Res Soc Develop 2020; 9: e3629108635.
[http://dx.doi.org/10.33448/rsd-v9i10.8635]

[171]   Bagavan A, Rahuman AA, Kaushik NK, Sahal D. *In vitro* antimalarial activity of medicinal plant extracts against *Plasmodium falciparum.* Parasitol Res 2011; 108(1): 15-22.
[http://dx.doi.org/10.1007/s00436-010-2034-4] [PMID: 20809417]

SUBJECT INDEX

A

Acetogenins 147, 160, 161
Acid 107, 141, 142, 143, 144, 148, 149, 150,
   151, 154, 155, 156, 159, 160
  arachidic 159
  araquidonic 156, 159
  behenic 156, 159
  betulinic 150, 151
  caprilic 156, 159
  carboxylic harman 142, 143, 144
  cativic 151
  cauranic-19-oic 151
  clorechinic 151
  copaiferic 151
  copalic 151
  ellagic 149
  estearic 156, 159
  folinic 107
  gallic 141, 148, 149
  hardwickiic 151
  hydroxycopalic 154, 155
  linoleic 156, 159
  myristic 156, 159
  oleic 156, 159
  palmitic 156, 159
  palmitoleic 156, 159
  phydroxybenzoic 160
  Pinifolic 155
  protocatechuic 148, 149
  ursolic 150
  vanillic 149
  vernolic 156, 159
Acidic phosphoprotein 20
Acinetobacter-derived cephalosporinases
  (ADC) 51
Acquired immune deficiency syndrome
  (AIDS) 14
Actagardine 87, 89
  naturally-occurring variant 87
Actinoplanes garbadinensis 87
Actinoplanes liguriae 87
Action 54, 136, 156

inhibiting 54
insecticide 136
population attributes healing 156
Activity 21, 22, 23, 24, 25, 28, 47, 54, 57, 80,
  86, 89, 108, 109, 110, 112, 113, 117,
  127, 141, 143, 147, 148, 149, 153, 154,
  158, 160, 161
  antibiotic 57
  antibiotic resistance 47
  anti-inflammatory 141
  bacteriostatic 80, 86
  endonuclease 28
  enzymatic 21, 22
  helicase 25, 26
  leishmanicidal 127, 143, 147, 154, 158
  narrow-spectrum 89
  polymerase 24
  protease 25
  restore beta-lactam 54
  reverse transcripitase 22
  toxic 149
  toxoplasmic 108
Acute pyelonephritis 61
Adjuvant therapy 91
Adolescents 76
ADP ribosyltransferase of rifampicin 52
Aedes aegypti 16
Agents 10, 22, 28, 47, 48, 91, 92, 111, 129,
  147, 161
  anti-inflammatory 111
  antileishmanial 147, 161
  antimalarial and antileishmanial 147, 161
  antimicrobial 47, 48
  developing anti-influenza 28
  etiological 129
  immunomodulatory 47
  leishmanicidal 161
  promising new anti-infective 92
  therapeutic 22, 91
Alkaloids 104, 110, 143, 144, 145, 157, 158,
  160, 161
  fraction of 143
  O-demethylaspidospermidine 145
Allergic disorders 6

www.ingramcontent.com/pod-product-compliance
Lightning Source LLC
Chambersburg PA
CBHW041702210326

41598CB00007B/509